Plasma Agriculture

Plasma Agriculture:

Oxygen Plasma Effects on Garlic

By

Matej Holc, Ita Junkar
and Miran Mozetič

Cambridge
Scholars
Publishing

Plasma Agriculture: Oxygen Plasma Effects on Garlic

By Matej Holc, Ita Junkar and Miran Mozetič

This book first published 2020

Cambridge Scholars Publishing

Lady Stephenson Library, Newcastle upon Tyne, NE6 2PA, UK

British Library Cataloguing in Publication Data
A catalogue record for this book is available from the British Library

ISBN (10): 1-5275-5815-0
ISBN (13): 978-1-5275-5815-1

TABLE OF CONTENTS

LIST OF ILLUSTRATIONS

List of Tables

PREFACE

This book examines the effects of gaseous plasma treatment on garlic cloves. The idea to expose garlic to the effects of plasma stems from the concept of "plasma agriculture", which has been a part of agricultural research trends for the last decade or so. Today, the physical phenomenon of non-thermal, highly reactive gaseous plasma is already widely employed in a variety of industrial applications, from microelectronics to medicine. The beneficial effects of plasma are mainly attributed to its unique mixture of reactive species, charged particles (ions, electrons), electric field, and ultraviolet radiation. In recent years, it has additionally become clear that, when properly applied, plasma can have a variety of beneficial effects on living organisms. In humans, this includes medicinal or cosmetic applications such as improved wound healing or effective teeth whitening. In the field of agriculture, however, many studies now show how exposure of seeds to plasma can boost plant germination, growth, or resistance of plants to disease. This includes important food crops such as wheat, corn, soy, and many more. These crops are all routinely propagated using seeds. Most garlic used in agriculture, however, is sterile and cannot produce seeds. Instead, it is propagated using cloves – the most well-known edible part of the plant, but also a vegetative propagation material, genetically identical to the parent plant, but simultaneously very different from the typical agricultural seed in size, weight, and water content.

By using garlic as a model organism to explain plasma agriculture principles, this work shows that many of them hold true for garlic cloves just as well as for seeds of agricultural plants. To illustrate this, it considers the physico-chemical effects of oxygen plasma on the garlic clove surface, together with the biological effects of such treatment. Plasma increases wettability of the clove surface; this is a result of functionalization, as the reactive oxygen species in the plasma oxidize the cuticle surface. Etching and the resulting nanostructuring of the surface additionally contribute to increased wettability. As a result of this hydrophilicity increase, water uptake of cloves increases as well. The biochemical mechanisms activated in this way accelerate clove germination and early growth of sprouts and roots. The garlic plants grown from optimally plasma-treated cloves also exhibit a higher yield than plants grown from untreated cloves. The behavior of unpeeled and peeled cloves differs due to surface chemistry and biological function differences.

The garlic used as sample material throughout this book is the Slovenian autochthonous cultivar *Ptujski spomladanski*. The improvements in growth and yield could strengthen its presence and availability, which are threatened by imported varieties. Thereby, local and sustainable agricultural practices could be supported by innovative technology practices. At the same time, the plasma technology is ecologically more acceptable than traditional, chemical-based approaches, as it uses abundant materials and does not introduce toxic compounds into the environment.

In addition to the specific effects of plasma on garlic, the book further seeks to explain the physical basics of low-pressure oxygen plasma. It stresses the importance of analyzing plasma characteristics directly, as opposed to simply adjusting discharge parameters. Non-thermal plasmas with conditions suitable for biological material treatment are commonly formed in laboratory environments, as well as scaled up to industrial-size reactors, both of which are used to generate the results presented in the course of this book. We show that by properly tuning the plasma and discharge parameters, as well as studying the physico-chemical effects of plasma on the treated material, it is possible to achieve the desired biological responses of the treated garlic cloves – that is, stimulation of garlic germination, growth, and improvement of its yield. As such, the book provides insights on the current understanding of plasma agriculture and encourages further steps in exploring the effects and benefits of this unique approach.

ACKNOWLEDGEMENTS

This work is an expanded adaptation of the doctoral dissertation titled "Effects of Highly Reactive Gaseous Plasma Treatment on the Germination and Growth of Garlic". The author of the dissertation, Dr. Matej Holc, would like to thank his supervisor, Prof. Dr. Miran Mozetič, and co-supervisor, Asst. Prof. Dr. Ita Junkar, at the Jožef Stefan Institute, Ljubljana, Slovenia. He would also like to thank the Jožef Stefan International Postgraduate School where the thesis was prepared and successfully defended.

The authors of this book would also like to thank Asst. Prof. Dr. Gregor Primc and Asst. Prof. Dr. Rok Zaplotnik, who have provided invaluable theoretical and practical support for the research activities involved. Prof. Dr. Janez Kovač and Prof. Dr. Alenka Vesel contributed important knowledge regarding surface analysis and processes. Tatjana Filipič and Janez Trtnik contributed excellent technical assistance.

Further thanks go to Dr. Jernej Iskra, Dr. Primož Titan, and Dr. Silva Grobelnik Mlakar, for their professional help and ideas. The team at the center for electron microscopy and microanalysis (CEMM) also offered immense help with all scanning electron microscopy analyses.

Special thanks go to Darko Vernik and the staff at Semenarna Ljubljana, Selection and Testing Center Ptuj, for their kindness and support, as well as for providing the knowledge and facilities for garlic field growth, harvest and analysis.

The authors further acknowledge the financial support from the Slovenian Research Agency ARRS, specifically Young Researcher Grant No. PR-06950 and research core funding grant No. P2-0082.

1. Garlic

1.1 Cultivation

Garlic (*Allium sativum* L.) is an important agricultural crop, with world production totaling over 24 million tons in 2013 (Food and Agriculture Organization of the United Nations 2015). Garlic growers aim to produce a high yield of a high-quality crop (Brewster 2008), but production remains limited by the region and climate (Wu et al. 2015, 43-52).

The main difference between the production of garlic and other crops is in the propagation. Cereals and other agriculturally cultivated plants are propagated sexually, through seeds. But while researchers have indeed discovered fertile clones of garlic some three decades ago in central Asia (Etoh 1986, 312-319), the commercially cultivated varieties of garlic remain sterile. As such, they do not produce seeds, and must be propagated vegetatively (Rabinowitch and Brewster 1989) – by planting individual cloves which germinate and produce new, genetically identical plants (Brewster 2008). This means that while it is possible to improve crops such as cereal, and even other alliums, by conventional or innovative breeding techniques (Mondal et al. 2016, 991), these are not applicable to agriculturally produced garlic, where we must consider alternative methods of crop improvement. Using such methods, we can improve garlic germination and growth without the need for genetic engineering, and propagate established or autochthonous cultivars instead.

Depending on climate and cultivar, garlic is typically planted either in the fall (September to November) or in the spring (January to March), both for a harvest in the summer. Ideally, the timing should be such that development begins after planting, but the plant does not emerge from the soil before the first frost (Kamenetsky 2007, 123-172). The local Slovenian garlic cultivars are either fall- or spring-planted. Among them, the cultivar *Ptujski spomladanski* was specifically adapted for long bulb storage time to enable spring planting. However, any garlic planted in the spring lacks the advantage of starting its development in the fall, and therefore requires early planting to properly develop its root system in time. Further, it may lack the low temperature exposure required to break dormancy and initiate clove formation and bulbing (Kamenetsky 2007, 123-172). External conditions such as wet soil or freezing of the soil due to low temperatures often hinder planting at the optimal time. To overcome these uncertainties brought on by external factors, it is beneficial to strengthen the garlic plant and improve its germination so that bulbs may grow efficiently, leading to high yields.

1.2 Anatomy

1.2.1 Bulbs and cloves

The underground part of a garlic plant is commonly called a bulb. Each bulb consists of several cloves. A clove is a storage bud consisting of a sprout leaf, found within a storage leaf, which is further covered by a protective leaf. The clove anatomy is schematically shown in Figure 1.1. The sprout leaf is the first leaf to emerge when the plant begins to actively grow, and is followed by foliage leaves. The storage leaf provides the food reserve required for the early growth. The protective leaf envelops the storage leaf of each clove, protecting it with its tough surface; it remains tightly wrapped around the clove when the bulb is broken apart. The cloves are further enveloped by layers of dry, white sheaths, which are the bases of the foliage leaves growing over ground. The anatomy of the protective leaf is very similar to that of the foliage leaf sheath (Mann 1952, 195-251). Leaves comprise of an upper and lower epidermis, and the mesophyll in between, with vascular bundles distributed throughout the latter (Evert and Eichorn 2012). In garlic, the epidermal cells are arranged in one row; a cross section of the top layers of the protective leaf is schematically shown in Figure 1.2. The storage leaf surface, which is the surface of the peeled clove, appears similar. However, its epidermis is comparably thin and delicate (Mann 1952, 195-251).

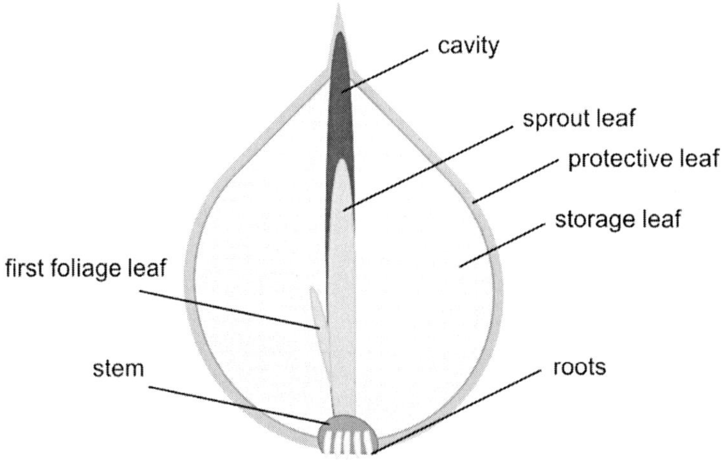

Figure 1.1: Garlic clove anatomy. Schematic representation. Adapted from (Mann 1952, 195-251).

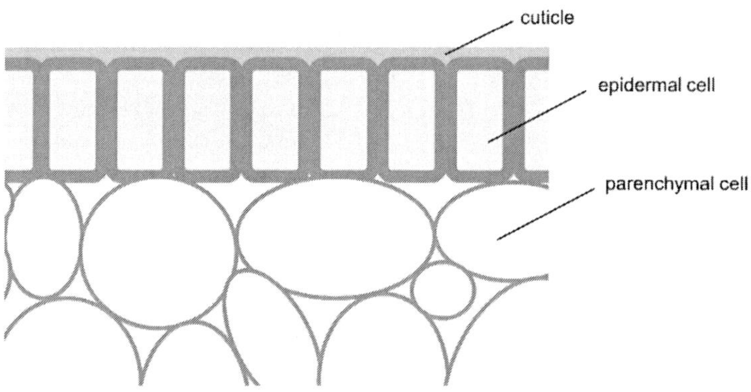

Figure 1.2: Schematic cross section of the protective leaf. Top layers are shown; the upper layer presents the sclerified epidermis, which is then further covered by a cuticle. Adapted from (Mann 1952, 195-251).

1.2.2 Clove surface chemistry

To understand how garlic cloves are affected by external treatments, such as exposure to gas plasma, we need to look at the composition of the very surface of clove components. The outer cellular layer of plant surfaces is the epidermis. As shown schematically in Figure 1.3, the epidermis is covered on the outer side by a cuticle (Mann 1952, 195-251), which is a layered, continuous non-cellular membrane composed of a polymeric lipid matrix embedded and covered with wax. In general, plant cuticle thickness is 0.1 – 10 μm (Riederer and Müller 2006). In garlic, this layer is known to be relatively thin (Mann 1952, 195-251).

Cuticular waxes are complex mixtures of compounds. As the name "wax" suggests, they consist of hydrophobic organic compounds. Chemically, the term wax usually refers to long-chain alkyl esters. However, in the cuticular context, it is used to also encompass compounds with different or additional functional groups, as well as other potential compounds present in this layer (Riederer and Müller 2006). Cuticular wax is additionally divided into two separate sub-layers: epicuticular waxes, which coat the outer surface of the cuticle, and intracuticular waxes, which are embedded in the cuticle matrix below. In a given plant species, chemical composition differs between the two sub-layers (Buschhaus and Jetter 2011, 841-853), but mainly in the relative amounts of the particular compounds (Riederer and Müller 2006). A standard array of cuticular wax compounds, found in virtually all plants studied to date, comprises fully saturated, unbranched hydrocarbon chains, from 20 to almost 40 carbon atoms in length, either as *n*-alkanes or containing a single terminal oxygen functional group: a hydroxyl, carbonyl, or carboxyl group, which gives a *n*-alcohol, *n*-aldehyde, or a fatty acid, respectively. In addition, the fatty acids and primary alcohols form an array of homologous esters, dimers of the aforementioned monomer compounds. The polymeric matrix of the cuticle is the biopolyester cutin, a primarily aliphatic polymer with long chains of methylene units linked by primary and secondary ester bonds (Riederer and Müller 2006). Glycerol is another monomeric component of cutin, and represents between 1 and 14 % of the total polymer composition (Graça et al. 2002, 205-215).

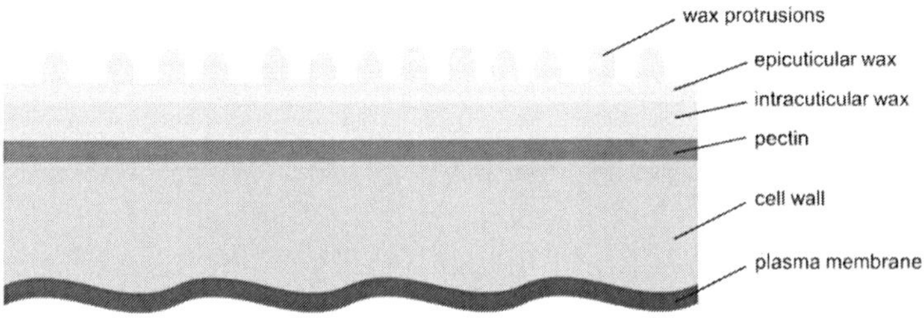

Figure 1.3: Anatomy of the outermost layers of plant epidermis. A pectin layer links the cuticle with the underlying cellulose cell wall of epidermis cells. Adapted from (Koch and Barthlott 2009, 1487-509).

1.2.3 Dormancy and germination

Germination is typically associated with emergence of the sprout from a seed. However, formally speaking, germination is the process from the first water uptake by the dry seed to the emergence of the radicle, which is the embryonic root of the plant. Further growth of the seedling, both sprout and root, after emergence of the radicle is called seedling establishment.

Before germination and growth may begin, plant propagation material of many species is in a resting state. Indeed, in a variety of plant seeds, as well as in garlic cloves, a phenomenon known as dormancy exists. The purpose of dormancy is to prevent germination in unfavorable seasons even if conditions are temporarily favorable. Dormancy needs to be released by additional environmental signals, such as light and temperature, before germination may begin. Release of dormancy allows for germination, which is the resumption of plant embryo growth after water uptake (Taiz et al. 2014; Graeber et al. 2012, 1769-86).

Garlic cloves are dormant when freshly harvested, and will not readily germinate (Mann 1952, 195-251). This period lasts for about 50 to 70 days, after which growth of the sprout leaf begins within the clove. As with seeds, dormancy of garlic cloves is controlled by the balance of plant hormones (Argüello et al. 1983, 1559-1563), and their external application to the clove accelerates germination (Rahman et al. 2006, 63-65). Roots emerge more rapidly than sprouts, even with dormancy breaking pre-treatment (Mann and Lewis 1956, 161-189). One important factor affecting garlic dormancy is temperature, as seed cloves need to be exposed to low temperatures for an optimal amount of time to achieve proper sprouting and bulbing. Normally this occurs after planting in the field, as garlic is planted in the autumn and harvested in the spring, thus experiencing the low temperatures during wintertime. Instead, cloves can also be exposed to low temperatures during storage, which affects both dormancy and bulbing (Brewster 2008).

2. PLASMA AGRICULTURE

2.1 Applications of plasma agriculture

Due to increasing food demands, and various factors affecting production, an improvement in the germination and growth of food crops is desired as a mechanism of increasing yields. Agricultural sustainability is desired to meet increasing food demands without harmful environmental effects (Araújo et al. 2016, 646). For example, wheat (*Triticum aestivum* L.) is one of the key cereal crops worldwide, and a major source of food. In 2015, an estimated 733 million tons of wheat were produced. In 2016, its required productivity growth was estimated at 1.1 % to meet consumption demands until 2020, while factors such as climate change, unpredictable environmental events, and various diseases or pests reduce yields and thereby further increase the need for production growth (Mondal et al. 2016, 991).

In agricultural research trends of the last decade, the discipline named plasma agriculture is a rapidly expanding field. It utilizes the processing capabilities of plasma treatment to improve properties of biological and food materials throughout the entire agricultural process, from seed to table. The aim of this research is to achieve optimal yields and increase food production, while simultaneously providing a high-quality, safe, environmentally friendly and affordable final product (Puač et al. 2018, 1700174). Plasma agriculture encompasses a variety of important applications such as disinfection of biological materials and food packaging materials, enhancement of seed or plant properties (Randeniya and de Groot 2015, 608-623), production and use of plasma-activated media (Thirumdas et al. 2018, 21-31), as well as plasma treatment of food (Thirumdas et al. 2015, 1-11).

2.2 Germination and growth improvement of agricultural seeds

In an attempt to avoid the excess use of fertilizers or chemical growth stimulants to improve plant performance in the field, plasma agriculture has delivered an environmentally acceptable alternative. The use of non-thermal plasma in agriculture is ecologically benign, does not introduce toxic compounds into the environment, and does not damage the sensitive planting material.

So far, gas plasma treatment of various agricultural seeds has been shown to affect many stages of plant germination and growth (Randeniya and de Groot 2015, 608-623). Germination rate, which is the percentage of germinated seeds at a given time point, has been improved for a variety of species, including seeds of quinoa (Gómez-Ramírez et al. 2017, 5924), rapeseed (Li et al. 2018), and peanut (Li et al. 2016, 1027-1033), using RF plasmas. After germination, various parameters of seedling growth were also improved, including sprout length, root length (Stolárik et al. 2015, 659-676; Dobrin et al. 2015, 255-260; Li et al. 2014, 5859), seedling dry weight (Stolárik et al. 2015, 659-676; Jiang et al. 2018, 044007), root fresh and dry weight (Henselová et al. 2012, 490-497), as well as other parameters. The selection of species included wheat, pea, maize, soybean, tomato, and other crops. A wide variety of plasma configurations has been used to achieve these effects, ranging from dielectric barrier discharge plasmas in atmospheric air to low pressure radio frequency discharges using a variety of process gases.

Further, planting of plasma-treated seeds can result in improved yield of agriculturally important crops. Increase in wheat yield was achieved by use of low-pressure helium plasma (Jiang et al. 2014, 54-58), low pressure air or air/helium plasma (Saberi et al. 2018, 11655; Zhang et al. 2018, 37-42), medium pressure air and air/oxygen plasma (Roy et al. 2017, 13-28), and an atmospheric pressure gliding arc discharge (Roy et al. 2018). Tomato yield was improved by treating seeds in an atmospheric pressure air plasma, either an arc discharge (Yin et al. 2005, 3143-3147) or a dielectric barrier discharge (Zhou et al. 2011, 23-27). A low pressure helium plasma improved peanut (Li et al. 2016, 1027-1033) as well as rapeseed yield (Li et al. 2018), while the seed yield of *Arabidopsis thailana* was improved by using air in an atmospheric pressure dielectric barrier discharge (Koga et al. 2016, 3).

2.3 Mechanisms of germination and growth improvement

Published research on plasma-induced seed germination and growth improvement currently focuses on the elucidation of physico-chemical and biochemical mechanisms by which plasma exerts its influence on the seeds. Plasma is uniquely suitable for modifying the outer layer of the seed surface, influencing its physico-chemical properties, without disrupting the bulk of the seed. As with polymer materials, the organic materials of the seed surface are targets for functionalization and etching by the reactive plasma species. In particular, treatment with a plasma that uses an O- or N-containing working gas increases the density of polar groups on the seed surface (Gómez-Ramírez et al. 2017, 5924; Bormashenko et al. 2012, 741). At the same time, plasma etching affects seed surface morphology by increasing its roughness (Meng et al. 2017, 1105-1119; Wang et al. 2017, 5601). Both increase the seed surface energy, causing hydrophilization (Zahoranová et al. 2018, 969-988). There appears to be a direct correlation between seed wettability and water uptake (da Silva et al. 2017, 280-285; Ambrico et al. 2017, 305401), which is the necessary first step in the seed germination process (Taiz et al. 2014). Increased hydrophilicity of the surface is therefore desired to enhance or accelerate seed hydration. Additionally, effects of etching may further facilitate seed coat permeability by the formation of cracks and pores on the seed surface (Stolárik et al. 2015, 659-676; Chen et al. 2012, 74-79).

In addition to these physico-chemical surface effects, the exposure of the seed to the plasma environment appears to activate certain biochemical pathways, thereby triggering a biological response. While the molecular mechanism of seed dormancy and germination are reasonably well known (Graeber et al. 2012, 1769-86), research in the field of plasma agriculture has only recently begun to investigate the effects of plasma on the seed biochemistry. These effects may be either specific, affecting certain signaling pathways, or part of a non-specific stress response of the seed to plasma treatment (Hayashi et al. 2015, 06GD01). The accelerated water uptake is a biological factor in itself, as it enables cell component mobility, restarting the seed metabolism after a period of quiescence or dormancy (Taiz et al. 2014). Further, concentrations of plant hormones involved in germination and dormancy are affected by plasma treatment (Ji et al. 2016, 117-28; Pérez Piza et al. 2018, 82-91; Stolárik et al. 2015, 659-676), an effect which may be related to the redox status and reactive oxygen species (ROS) signaling pathways of the seed. ROS are proposed to act as transmitters of environmental cues during germination, and the concept of an "oxidative window for germination" has been proposed to describe a critical range of ROS content in seeds that allows for dormancy release and germination (Bailly et al. 2008, 806-814). Exposure to the reactive plasma species could thus affect these mechanisms, and thereby, germination and growth.

By all of the above, plasma treatment of seeds can be considered a physical seed priming approach. Effect of plasma on both germination and yield varies with the selected plasma parameters, as well as plant species, and should be optimized for each specific treatment case.

3. GASEOUS PLASMA

3.1 Fourth state of matter

Gaseous plasma is one of the four fundamental states of matter, following the solid, liquid, and gaseous states. Adding sufficient additional energy to a gas results in a plasma – a state in which a substantial amount of molecules is dissociated and ionized, while the gas remains macroscopically neutral. The plasma state is abundant in nature. It is found on Earth in the form of lightning and aurora borealis, and comprises the majority of the remaining universe, such as the interior and corona of stars, or the solar wind (Fridman 2008). On the other hand, plasma is commonly formed in laboratory settings in various ways, typically by applying heat or a strong electric or electromagnetic field to a gas. Laboratory plasma systems vary widely in the possible selection of power sources, electrode configurations, reactor size, process gas, working pressure, etc. (Conrads and Schmidt 2000, 441-454). The purpose of such man-made plasmas is often tailoring of surface properties. Scaled up to an industrial level, they are applied in material, medical, electronic, and other sciences.

3.2 Thermal plasma

A plasma state achieved by gas heating is a thermal plasma. If a material is in a gaseous state, and the temperature of a gas is further increased, the molecules of the gas will dissociate into atoms, which at even higher temperatures will further dissociate into charged particles: electrons and ions. This effect is called ionization. Very high temperatures, in the range of several thousand Kelvin, are required for this to occur (Lieberman and Lichtenberg 2005). As such, thermal plasmas are unsuitable for direct treatment of biological materials or polymers, as at the required temperatures, thermal degradation of the treated material would take place. Thermal plasma is, however, a rich source of chemically reactive species such as atoms, radicals and metastable molecules. Thus, indirect treatment with thermal plasma may be of benefit, while also allowing to avoid substantial temperature increases and thermal degradation of treated materials.

For example, one hot plasma source of cold reactive gaseous species is a gliding arc discharge (Fridman et al. 1999, 211-231). The device is schematically presented in Figure 3.1. The distance between its electrodes is gradually increasing. There is a pressure difference across the device, which causes drifting of the gas through the narrowest gap between the electrodes and the gas expansion thereafter. The discharge is ignited by a simple DC power supply connected to one electrode, while the counter-electrode is grounded. This discharge, ignited at the narrowest part between the electrodes, causes formation of an electric arc (a hot plasma) in the gap between the electrodes. The gas flow pushes the arc towards the wider gap between the electrodes, where the gliding arc is observed. The gas temperature in the gliding arc is much lower than in the original arc in the narrowest gap, but charged particles as well as radicals persist even far away from the narrowest gap. The materials to be treated by the reactive gaseous species are placed at the widest gap between the electrodes, where the plasma is not thermal any more, but still rich in reactive gaseous species capable of interacting chemically with organic matter.

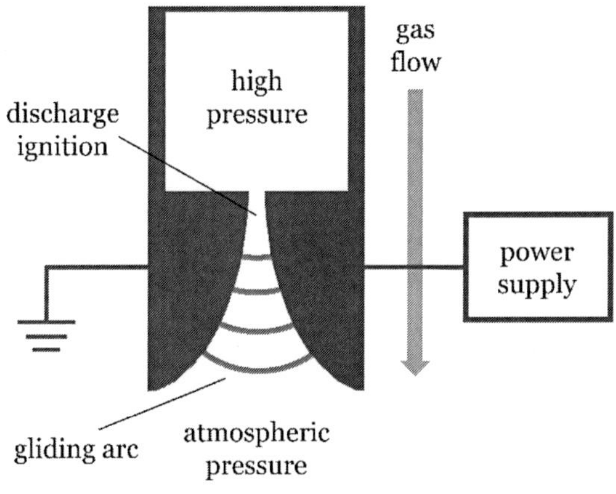

Figure 3.1: Gliding arc discharge. Schematic representation.

An alternative to the gliding arc is application of a hot plasma which expands into a vacuum chamber (Engeln et al. 2002, A100-A104). Figure 3.2 shows such a plasma expansion configuration schematically. There is a narrow nozzle between a dielectric tube and a vacuum chamber. The gas flows through the dielectric tube and expands into the vacuum chamber. Hot gaseous plasma is sustained in the dielectric tube with a suitable electric discharge, while inside the vacuum chamber, there is an expanding plasma whose temperature decreases rapidly as the gas expands into the vacuum chamber. The configuration in Figure 3.2 is similar to that in Figure 3.1, except that the vacuum conditions inside the vacuum chamber suppress the loss of chemically reactive species in the gas phase. Namely, the life-time of reactive species depends inversely on the square of the pressure. The speed of gas expanding into the vacuum chamber reaches the sound velocity at the nozzle, so the transport of reactive plasma species from the hot plasma to the vacuum chamber is extremely efficient.

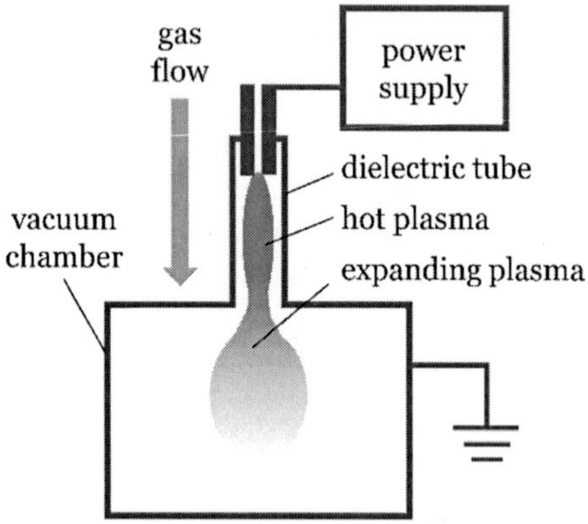

Figure 3.2: Plasma expansion. Schematic representation.

3.3 Non-thermal plasma

In contrast to thermal plasmas, non-thermal plasmas are formed at temperatures near room temperature. They can be initiated by several types of discharges, and at a wide range of pressures. The most common distinction is between atmospheric pressure and low-pressure discharges; both are considered in detail in the following chapters. The discharges are electrically driven, and the primary recipients of the energy transfer are electrons, which are mobile and charged (Lieberman and Lichtenberg 2005). This results in a pronounced inequality between the kinetic temperatures of electrons and the heavy particles (ions, neutral particles) in the plasma. The electron temperature is much larger, typically several 10,000 K, while the temperature of other species remains close to room temperature. Furthermore, the concentration of reactive species in non-equilibrium plasma is much larger than the value one can calculate based on room temperature using Boltzmann distribution. Therefore, the plasma is not in thermodynamic equilibrium, and cannot be described by the basic laws of equilibrium thermodynamics (Gorjanc and Mozetič 2014). In this environment, numerous reactions occur between the plasma particles. Electron collisions with atoms and molecules bring on processes such as dissociation, excitation and ionization, which results in the occurrence of heavier particles with relatively high potential energy (Lieberman and Lichtenberg 2005).

In fact, one should be cautious when using the term "temperature" while describing non-equilibrium gaseous plasma. Strictly, temperature is not defined for non-equilibrium systems, but in practice, it is used to describe the average kinetic energy of gaseous species which assume the Maxwell-Boltzmann distribution. In scientific literature, various temperatures have been adopted for description of non-equilibrium gases: the neutral gas kinetic temperature, the electron temperature, the ion temperature. All these temperatures indicate the distribution of corresponding particles over kinetic energy (Lieberman and Lichtenberg 2005). The distribution over excited states is rarely referred to as "temperature", although the terms "vibrational temperature" or "rotational temperature" have been adopted (Britun et al. 2007, 1022-1029). Most authors mean "neutral gas kinetic temperature" when using the term "temperature" to describe non-equilibrium gases. This was also adopted in the present monograph, unless stated otherwise.

3.4 Atmospheric pressure plasma

3.4.1 Collision phenomena

Non-equilibrium gaseous plasma can be sustained in different gases at different pressures. The voltage between the two electrodes should be high enough to enable acceleration of free electrons, where as a result they gain enough energy to cause ionization, dissociation, or excitation of gaseous molecules upon inelastic collisions. A high concentration of charged particles as well as atoms or molecules in highly excited states at low gas temperature is against the rules of thermodynamics, so the reactive species tend to neutralize, relax, or associate to stable molecules (Fridman 2008). The concentration of reactive gaseous species obviously depends on the production and loss rates. The production depends of various parameters, but the major one is the applied discharge power. A higher power will generally lead to a higher concentration of the reactive gaseous species (Zaplotnik et al. 2012, 3857-67).

There are numerous channels for the loss of reactive species, some of them taking place in the gas phase and some on the surfaces of any material facing plasma. A collision of an electron and a positively charged ion in the gas phase usually does not lead to neutralization. The reason is the requirement for conservation of energy and momentum. Before the collision, there is high potential energy in the electron–positive ion system. This potential energy is equal to the ionization energy of the parent molecule, which is often above 10 eV (National Institute of Standards and Technology 2019). If neutralization occurs, the potential energy prior to the collision should appear as the energy of the newly formed molecule. The requirement for conservation of momentum prevents the potential energy before the collision to appear as the kinetic energy of the molecule. It can appear only as the internal energy of the neutral molecule. Since the potential energy before the collision is large, the large internal energy would lead to immediate dissociation, a reaction that is regarded useful for plasma chemistry. Still, the probability for such dissociative neutralization in the gas phase is low.

Similar considerations should be taken into account for the case of association of two atoms to the parent molecule. The potential energy before the collision between two neutral atoms in the ground state is equal to the dissociation energy of the parent molecule. The potential energy before the association in the gas phase cannot be transferred to the kinetic energy of the newly formed molecule due to the requirement for the conservation of momentum. The atoms are therefore rather stable in the gas phase, despite numerous collisions among them.

An effective channel of dissipation of the potential energy of colliding particles in the gas phase are three-body collisions. Suppose a simultaneous collision of three particles, two atoms and a molecule, occurs. The collision proceeds as: atom + atom + molecule → two molecules. These resulting two molecules easily take the potential energy before the collision in the form of kinetic energy, which is shared equally between the molecule involved in the collision and the newly formed molecule (Lieberman and Lichtenberg 2005). In the case of three-body collision, the momentum is conserved, since the two molecules simply move apart after the collision. The kinetic energies of the two molecules after the collision should obviously be equal to the potential energy before the collision. Such a collision therefore ensures conservation of energy and momentum, and so such an association of two atoms to the parent molecule is likely to occur as long as the frequency of three-body collisions is large.

The probability that three particles meet at the same time in the same place depends on the mean free path of gaseous particles between subsequent collisions, which in turn depends on the density of gaseous particles and their speed of thermal motion (Misra et al. 2017, 1832-1863). The higher the gas temperature, the more frequent the collisions. The density of gaseous molecules depends on the pressure. The standard equation is $p = n\, k_B\, T$, where p is the gas pressure, n the density of gaseous particles, k_B the Boltzmann constant, and T the gas temperature (measured in K). The collision frequency therefore increases with increasing pressure.

Of particular importance are three-body collisions that lead to association of atoms to parent molecules. The frequency of three-body collisions increases with the square of the pressure. At atmospheric pressure, the frequency is roughly about 1 MHz, while at the pressure of 1 mbar, it is solely about 1 Hz. There are about one million collisions per second that may lead to the loss of reactive plasma particles at atmospheric pressure. The loss of such particles at atmospheric pressure is thus much more intense than at low pressure (Zaplotnik et al. 2018, 1800021). This is the major reason for choosing low-pressure plasma for treatment of various materials on the industrial scale.

3.4.2 Heating

The loss of reactive plasma particles by three-body collisions causes localized heating. As explained above, the potential energy before the three-body collision is transferred to the kinetic energy of the molecules (Lieberman and Lichtenberg 2005). The original and newly formed molecules therefore move apart at rather high kinetic energy. The dissociation energy of a gaseous molecule is usually several eV. For example, the dissociation energy of the oxygen molecule is 5.2 eV (Luo and Kerr 2012). This energy is shared between the original and newly formed molecule equally, so the kinetic energy of each molecule after the three body collision is about 2.6 eV. An energy of 1 eV corresponds to the temperature of about 11,000 K. The molecules are therefore super-fast, or super-thermal, after the three-body collision. This is why such collisions are often called super-elastic collisions.

After a super-elastic collision, the molecules collide with other (thermal) molecules. The collisions are often elastic: the molecules keep their original internal energy (in view of vibrational and rotational states), but a significant fraction of the kinetic energy of the fast molecule is transferred to the slow molecule (Lieberman and Lichtenberg 2005). The fast molecules formed at super-elastic collisions therefore quickly lose their energy and the rest of gaseous molecules gain the energy. The energy gained from the super-fast molecules therefore causes heating of the gas – an increase of the neutral gas kinetic temperature.

Not only fast molecules formed upon super-elastic collisions, but also other reactive gaseous particles in plasma share their energy with thermal molecules. For example, molecules that have been excited to electronically excited states will de-excite upon collisions with other gaseous molecules, and the potential energy is often shared between the colliding particles in view of the kinetic energy. As explained above, the collisions are numerous at atmospheric pressure, so the gas is heated extensively.

There are methods for suppressing extensive gas heating upon plasma conditions at atmospheric pressure. One method was already explained: drifting gas (typically at room temperature) through gaseous plasma. As long as the discharge power is rather low and the gas speed high, the heating of the gas passing plasma is tolerable.

In any case, practically all power spent for sustaining the gaseous plasma at atmospheric pressure is transformed to the increases in gas temperature. One can estimate the temperature of gas passing an atmospheric pressure plasma from a simple equation: $\Delta T = P \,/\, (\varphi_V\, \rho\, c)$. Here, ΔT is the temperature difference between the gas before and after passing plasma, P the discharge power, φ_V the gas flow rate, ρ the gas density, and c the gas specific thermal capacity. For air at ambient conditions, the density is roughly 1 kg/m^3, and the specific thermal capacity is roughly 1000 J/kg K. In Figure 3.3, the temperature difference is plotted versus the discharge power for two different gas flows often used in experiments with plasma jets, i.e., 1 and 10 standard liters per minute (slm).

The results of the simple calculation indicate that gas heating in atmospheric pressure plasma may not always be neglected. For example, air passing plasma sustained by a power of only 1 W at the flow of 1 slm will be heated for 60 K. If the original gas is at room temperature (about 20 °C), the gas will heat to 80 °C. The gas will heat for only 6 K if the air flow is 10 slm and the discharge power remains 1 W. Many plasmas sustained in air jets are powered with much larger powers. In such cases, the gas heating may be prohibitively high.

These simple calculations, summarized in Figure 3.3, assume that all discharge power is spent for heating the gas, neglecting any cooling. The gaseous plasma glows, so part of the discharge energy is always emitted by radiation (Cvelbar et al. 2007, 224-227). This effect is negligible at low powers, but becomes important at low flow rates and/or large discharge powers. The emission appears in a range of wavelengths and arises from radiation transitions from high to low excited states (sometimes to the ground state) (Lieberman and Lichtenberg 2005). Plasma at high temperature (although non-equilibrium) also radiates a continuum as a black body.

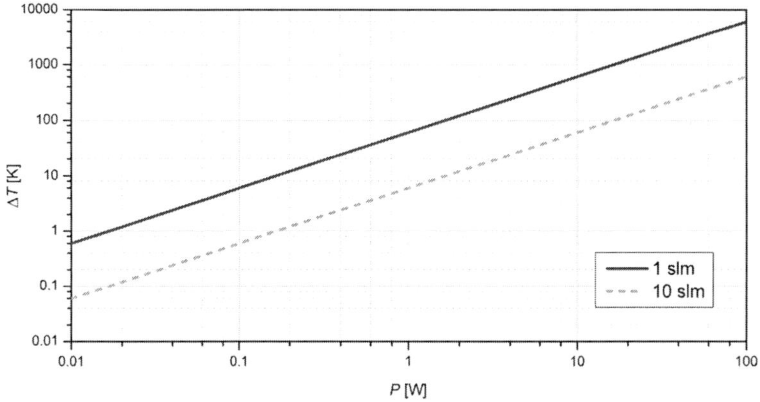

Figure 3.3: Temperature difference versus discharge power. Estimation for an atmospheric pressure plasma.

The heating of gas passing an atmospheric pressure plasma is due to absorbed power. The power is not always trivial to measure. The first approximation is simply the power spent by the power supply. This power, however, is always higher than the power absorbed by gaseous plasma. One should therefore be cautious when stating the discharge powers.

One method for suppressing gas heating by three-body collisions is application of gaseous plasma sustained in almost pure noble gases. Argon as the most abundant noble gas and helium as the second one are both popular. Reactive gases like air, oxygen, water vapor, etc., are added in minute quantities. It has been shown by numerous authors that 1 vol.% of reactive gas in the noble gas is more than enough to obtain an appreciable concentration of reactive gaseous species such as O or N atoms, NO_x molecules, and similar (Li et al. 2009, 111501). The key advantage of using noble gases at atmospheric pressure is the ability to sustain the plasma at a low discharge power. The reason is a lack of three-body collisions that would lead to super-elastic collisions and thus heating of the gas upon passing the plasma zone. For example, the collision between two Ar atoms and an O atom will have no effect on the plasma properties; collisions will simply be elastic. Furthermore, noble gases are known for their metastable states (Uchida et al. 1980, 881-883). A fast electron will excite an Ar atom to a metastable state, and thus give potential energy to the Ar atom. The atom will remain excited until a collision with an oxygen molecule occurs.

At such a collision, the dissociation of the oxygen molecule to parent atoms occurs, as the excitation energy of the Ar metastable is larger than the dissociation energy of the oxygen molecule. The metastable states of noble gases therefore act as a reservoir of energy for excitation of reactive gases to useful species.

Plasma sustained in noble gases is a standard in scientific research on the interaction between reactive gaseous species and organic material. Such plasma can be sustained at low discharge power, so gas heating is often negligible. Furthermore, the plasma is more uniform and the gradients of reactive plasma species smaller than in plasmas sustained in reactive gases themselves, such as air, oxygen, water vapor, etc. Atmospheric pressure plasma sustained in argon may be also a rather intense source of vacuum ultra-violet (VUV) radiation (Brandenburg et al. 2009, 877-883). Such radiation involves photons of energy well above the binding energy between atoms in organic matter, which therefore cause modification of such materials. The penetration depth of VUV radiation depends on particular energy of the photons, but as a rule of thumb, it is roughly 30 nm at a photon energy of 10 eV. The penetration depth of such radiation in air is well below a millimeter.

3.4.3 Atmospheric pressure plasma configurations

One configuration of an atmospheric pressure plasma is a plasma jet, schematically shown in Figure 3.4 (Schutze et al. 1998, 1685-1694). Such plasmas are usually sustained in a noble gas for reasons explained above. Typically, the gas flows through a dielectric tube. There is an electrode in the tube, connected to a power supply that often operates at voltages up to about 10 kV, powers of a few W, and frequencies of several kHz. Numerous streamers expand from the electrode in the direction of the gas jet, making an impression of continuous plasma within the gas jet. The visible plasma expands as far from the electrode as the electric field is capable of sustaining the electron streamers and as far as the noble gas does not mix with ambient atmosphere.

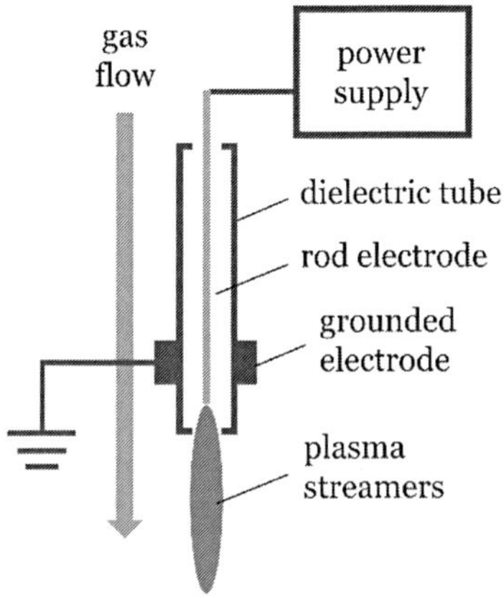

Figure 3.4: An atmospheric pressure plasma jet. Schematic representation.

Atmospheric plasma jets sustained in air are often powered by radio-frequency (RF) (Zhang et al. 2013, 205202) or even microwave (MW) discharges (Kim et al. 2008, 191505), i.e., discharges in the frequency range between several MHz and several GHz. An example of such a configuration is shown in Figure 3.9. Plasma sustained at such frequencies is continuous – no streamers are observed due to the large frequencies. The gas is never cold in such discharges, and the temperature is easily estimated from the discharge power – see Figure 3.3.

Yet another popular configuration of the atmospheric pressure non-equilibrium gaseous plasma involves planar discharges, often called dielectric barrier discharge (DBDs). Figure 3.5 shows a schematic of such a device. DBD plasma systems are set up as two electrodes with a dielectric barrier and a thin gap, such as a few millimeters, in between. A large number of plasma filaments is formed in the discharge (Conrads and Schmidt 2000, 441-454). Unlike the plasma jet, where the plasma is rather one-dimensional, the planar configuration enables two-dimensional plasma, which may be useful for treatment of three-dimensional objects. Co-planar discharges have been invented as well. In such cases, there is a grid of electrical conductors on one side of a dielectric plate while the gaseous plasma appears on the other side, providing the conductors are embedded into perfectly non-conductive medium, such as non-polar liquids (Dobrin et al. 2015, 255-260; Henselová et al. 2012, 490-497).

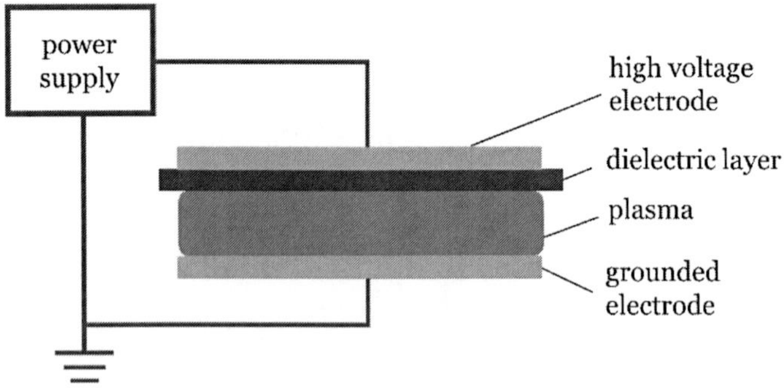

Figure 3.5: A planar atmospheric plasma configuration. Schematic representation.

While atmospheric pressure plasmas are easy to use, they are rather difficult to characterize. As already stressed, the abundance of three-body collisions causes rapid loss of reactive gaseous species. A significant concentration of such species at atmospheric pressure is thus assured only in the volume where the electrons are accelerated, i.e., in the volume of high electric field. Outside such a volume, the reactive plasma particles are lost by gas phase collisions. The atmospheric pressure plasmas are therefore known for huge gradients in the density of reactive species, which makes uniform treatment of three-dimensional objects virtually impossible. Some recently published papers on characterization of atmospheric plasma sources include: (Golda et al. 2020, 1900216; Kehrer et al. 2020, 125170; Luan et al. 2019, 031305; Nishime et al. 2020, 1028). In industry applications, robots are often used to manipulate the objects to be treated by atmospheric pressure plasmas.

3.5 Low-pressure plasma

3.5.1 Vacuum chamber evacuation

The disadvantages of atmospheric pressure plasmas are overcome by using low-pressure plasmas. In such cases, the gradients in reactive plasma particles are almost absent and the uniform plasma expands in large volumes, often measured in liters, if not cubic meters. Such plasmas are very popular in industrial applications. Different pressures are used in different applications, but almost all low-pressure technologies for tailoring surface properties of solid materials employ plasma sustained in the pressure range between 1 Pa and 1 mbar. The definition of low-pressure plasma, as adopted in this book, refers to this range of pressures unless stated otherwise.

Basic concepts of collision phenomena described above are, of course, valid also for low-pressure plasma, but one should take into account that the collision frequency in the range of pressures useful for sustaining low-pressure plasma is so low that any loss or reactive species in the gas phase is marginal. As a consequence, the discussion about the gas temperature and the results presented in Figure 3.3 are void for low-pressure plasma. At low pressures, the discharge power is rarely used for heating the gas, but predominantly heats the surfaces of materials facing the plasma.

Low-pressure non-equilibrium gaseous plasma is sustained in hermetically tight vacuum chambers. The vacuum chamber is pumped with an appropriate vacuum pump (Bekkara et al. 2017, 225-231). Once the pump is turned on, the pressure in the vacuum chamber begins to decrease as the gas inside the chamber is slowly removed from the chamber. The removal rate depends on the volume of the vacuum chamber, the pumping speed of the vacuum pump, and the conductivity of any elements placed between the vacuum chamber and the pump. The higher the pumping speed and conductivity of any element mounted between the chamber and the pump, the lower the pressure in the vacuum chamber that can be achieved (Hoffman et al. 1998). The time dependence of the pressure in a hermetically tight vacuum chamber is shown in Figure 3.6.

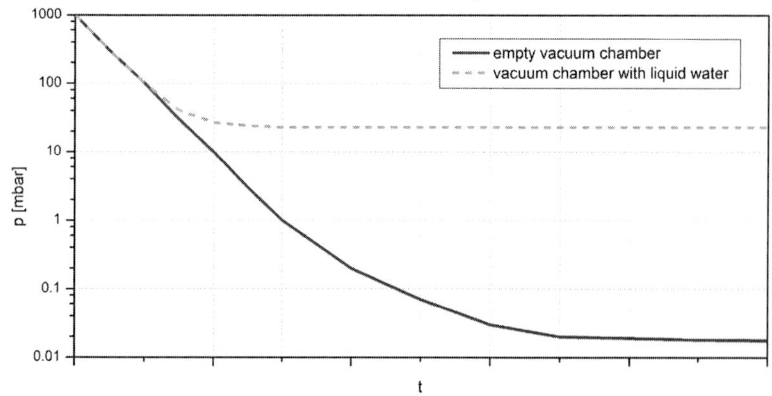

Figure 3.6: Pumping curves of a vacuum chamber. Empty vacuum chamber (solid curve); vacuum chamber with liquid water kept at a constant temperature of 20 °C in the vacuum chamber (dashed curve).

The pressure decreases rather exponentially at first, but after achieving a certain vacuum level, a deviation from the exponential decrease is observed. The deviation is due to various reasons. For example, the conductivity of any element between the vacuum chamber and the pump decreases with decreasing pressure. Another reason may be the deviation of the pumping speed from a constant value at low pressure. Vacuum pumps are designed to evacuate chambers only in a range of pressures. When the pressure approaches the ultimate pressure of the vacuum pump, the pumping speed approaches zero. There are several reasons for this effect, including the final compression rate, the vapor pressure of the liquid used for sealing the pump, and back-streaming of gaseous molecules (Hoffman et al. 1998). Another important reason may be desorption of gaseous molecules from the walls of the discharge chamber. Gaseous molecules tend to adsorb on surfaces and are bound there by attractive forces such as Van der Waals interactions. The adsorbed molecules slowly leave the surface. The desorption rate depends on the surface temperature. Of particular importance are water molecules which are adsorbed at the surfaces of any material placed inside the vacuum chamber. They abound on the surface and desorb slowly upon evacuation. Other molecules such as nitrogen, oxygen, argon and carbon dioxide (main constituents of dry air) desorb much faster, so their contribution to desorption upon prolonged evacuation is marginal compared to water molecules (Caswell 1960, 130-142). The consequence of such slow desorption of water molecules from the inner surfaces of vacuum chamber is that the residual atmosphere in the hermetically tight vacuum chamber consists practically only of water vapor. Any treatment of materials with gaseous plasma sustained in a selected gas intentionally introduced into the vacuum chamber should take into account that plasma is not sustained in the selected gas alone, but rather in a mixture of the selected gas and water vapor. It is possible to remove the residual water by pumping the chamber with pumps of high pumping speed and very low ultimate pressure during continuous baking – i.e., heating of the entire vacuum system to elevated temperature, for example 90 °C. Such a procedure is impractical and obsolete for the case of treatment of biological material, as will be illustrated in the proceeding text.

Suppose there is a container of water in a vacuum system. The liquid water does not affect the pumping curve as long as the pressure in the vacuum chamber is above the water vapor saturated pressure. The temperature dependence of the water vapor saturated pressure is shown in Figure 3.7. The saturated vapor pressure at a given temperature reflects the boiling point. At the atmospheric pressure, i.e., 1 bar, the boiling point is 100 °C. As long as the temperature is below the boiling point, water slowly desorbs from the liquid, and the partial pressure of water vapor increases with time until the saturated pressure is achieved. Further on nothing really happens since the amount of water evaporating from the liquid equilibrates with the amount of water condensing. The boiling point decreases with decreasing pressure, as illustrated in Figure 3.7. Upon evacuation of the chamber, the pressure decreases until the saturated vapor pressure is achieved. Any further pumping will result in boiling of the liquid water. Boiling is an extremely effective method of forming vapor, much more so than evaporation. The consequence is an inability to obtain vacuum levels below the saturated vapor pressure in a chamber that contains liquid water. Even very high pumping speeds of the vacuum pump do not allow for evacuation of the vacuum chamber below the saturated vapor pressure.

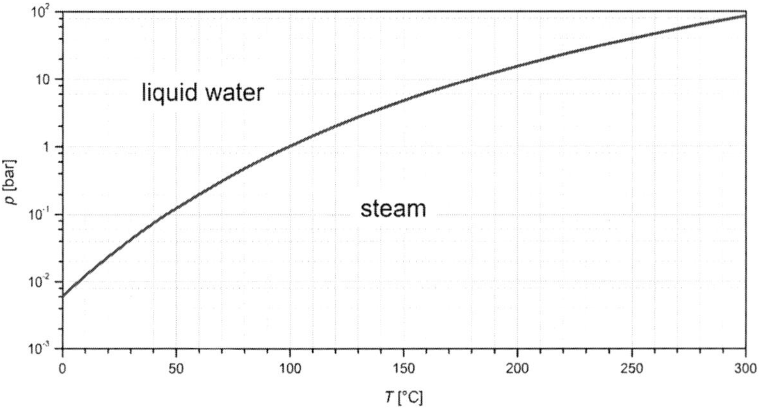

Figure 3.7: Water vapor saturated pressure versus temperature. Data from (Engineering ToolBox 2004).

The pumping curve for the case of liquid water kept in the vacuum system is shown as a dashed curve in Figure 3.6. Once the saturated vapor pressure is achieved, the curve is a perfect constant. Such a constant value is observed only in case the water is kept at a constant temperature. The effect of boiling causes cooling of the boiling water. Maintaining constant temperature is thus possible only through heating of the boiling water. If the water is not heated, its temperature will decrease with increasing pumping, while the pressure inside the vacuum chamber will also decrease, as shown in Figure 3.7. Eventually, the water will freeze. Even if ice is present in the vacuum chamber, the vacuum level will not be perfect, since the vapor pressure at 0 °C is still 6 mbar, as revealed in Figure 3.7. Prolonged evacuation of the chamber containing a piece of ice will cause further decrease of the ice temperature.

These facts influence the achievable vacuum level in a chamber filled with water-containing materials such as garlic. A clove of garlic contains about 65 % of water (Kamenetsky 2007, 123-172). However, this water is bound, so the desorption behavior will not be the same as when a container of liquid water is placed into the vacuum chamber. Instead, the pumping curve will be somewhere between the solid and dashed curves of Figure 3.6. The achievable vacuum level obviously depends on the pumping speed of the vacuum pump, as well as the amount and properties of the biological matter placed inside the chamber. The biological matter will also cool upon vacuum conditions. A small amount of garlic placed inside the vacuum chamber will cause only moderate desorption, so the achievable vacuum level is good enough to sustain a low-pressure plasma in the chamber containing biological matter.

3.5.2 Breakdown voltage

A properly evacuated chamber is ready for ignition and sustaining of a gaseous plasma. The electric field applied across the gas should be above a critical level, which should be determined experimentally. The voltage needed for the gas breakdown depends on the pressure inside the vacuum chamber, the distance between the electrodes, as well as on the type of gas or gas mixture. The breakdown was found to depend on the product of the distance between the electrodes and the pressure in the vacuum chamber. The dependence of the breakdown voltage versus the product of the distance and pressure is known as Paschen's law (Fridman 2008). Figure 3.8 shows such a dependence. A minimum in the curve is observed. It depends slightly on type of the gas, but as a rule of thumb, the minimum occurs at the product value of about 1 cm mbar. At lower product values, the breakdown voltage increases rather steeply with decreasing product value, whereas at product values higher than the minimum, the curve increase is not that steep. The Paschen curves reflect a simple fact: it is difficult, if not impossible, to ignite a plasma at elevated pressures in large plasma chambers. On the other hand, it is also difficult to ignite a plasma in small chambers at very low pressure. Usually, the limiting factor is at product values higher than the minimum in the Paschen curve. For example, if one wants to ignite plasma at the water vapor saturated pressure at 20 °C (23 mbar, see Figure 3.6), the optimal conditions are roughly at a distance between the electrodes as small as 0.4 mm. This is not very practical for treatment of macroscopic samples, so a pressure well below the saturated water vapor pressure should be achievable in a vacuum chamber for treatment of garlic cloves. As explained above, presence of liquid water in the vacuum chamber should be avoided in order to be able to ignite gaseous plasma in a chamber of reasonable dimensions. For example, Figure 3.8 reveals that in a chamber with an inter-electrode distance of 10 cm, the optimal pressure is 10 Pa. Such a low pressure may not always be achievable. Happily enough, the increase of the breakdown voltage with increasing value of the product of pressure and distance between the electrodes is not dramatic, so one can ignite plasma at 10 cm inter-electrode distance even at the pressure of 1 mbar, provided the voltage of the power supply is at least 1000 V.

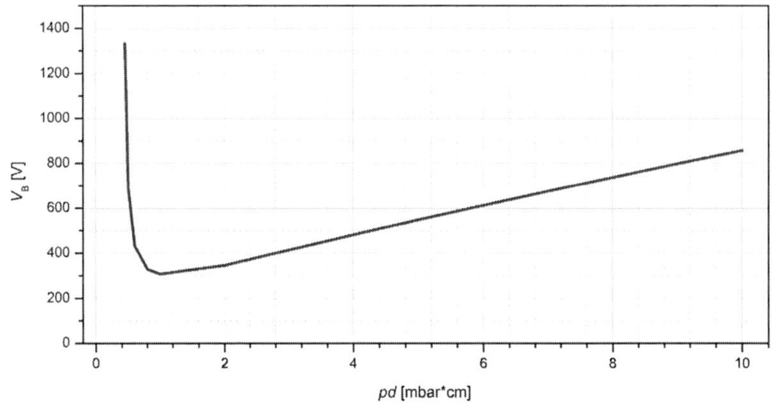

Figure 3.8: Paschen's law. Example for air. Data from (Bazelyan and Raizer 1997).

3.5.3. Electrical properties

By definition, gaseous plasma contains free electrons and positively charged molecular ions. Some plasmas, for example oxygen plasma, also contain a significant amount of negatively charged molecular ions, which are formed upon attachment of slow electrons to a neutral molecule. The negatively charged O_2^- molecules are stable in the gas phase (Lieberman and Lichtenberg 2005). The mobility of charged particles depends on their mass: a lighter electron will have a much larger mobility that an ion, which has a mass a thousand times (or more, depending on the molecular mass) larger. As a consequence, the fast electrons will reach the surface of any material facing plasma much faster that the positive ions. The surface will therefore be charged negatively against the gaseous plasma. The negative potential on the surface will reflect most electrons arriving towards the surface, and on the other hand, accelerate positively charged ions (Vesel and Mozetič 2017). The potential between the gaseous plasma and the surface is self-adjusted according to the mobility of electrons and positive ions, so that the fluxes of both types of charged particles to the surface of a material facing plasma will equalize. This happens very quickly after ignition of the gaseous plasma, typically on a time scale below a microsecond. The potential between the gaseous plasma and the surface depends on the mobilities of both types of charged particles, i.e., the electron temperature in the gaseous plasma, as well as the mass of the positively charged ions (Fridman 2008). For oxygen plasma of low discharge power, which is useful for treatment of biological materials, the potential difference is about 10 V. This voltage will therefore equilibrate the fluxes of positively and negatively charged particles onto the surface. Only the electrons from the high-energy tail of the electron energy distribution function will be able to reach the surface. In practice, this means only a few percent of all plasma electrons. The negatively charged ions are also reflected from the negatively charged surface, so they will never be able to reach the surfaces. This is why their density in electronegative gases may be substantial.

Gaseous plasma is an electrically conductive medium. The electric field originally applied to gas (before the breakdown) will not be able to penetrate the gaseous plasma, because electric fields cannot penetrate conductive materials. As a consequence, the constant field between the electrodes collapses as soon as gaseous plasma appears in the volume between the electrodes. The electric field is rather concentrated next to the electrodes. Since the mobility of electrons is large, the field does not concentrate next to the anode, but only next to the cathode. The electric field in the sheath next to the cathode obviously increases with increasing density of free electrons in the gaseous plasma. The thickness of the sheath between the gaseous plasma and the cathode is often on the order of 10 Debye lengths. The Debye length for highly non-equilibrium gaseous plasma has been defined as

$$\lambda_{\mathrm{D}} = \sqrt{\frac{\varepsilon_0 k_{\mathrm{B}} T_e}{n_e q_e^2}}, \tag{3.1}$$

where λ_{D} is the Debye length, ε_0 is the permittivity of free space, k_{B} is the Boltzmann constant, T_e is the electron temperature, n_e is the density of electrons, and q_e is the charge of an electron (Hutchinson 2002).

The voltage across the sheath between the cathode and bulk plasma is rather close to the voltage of the DC power supply, i.e., several 100 V. This voltage is large enough to reflect all electrons coming from the gaseous plasma. The positively charged ions are accelerated towards the cathode. If the sheath is almost collision-less (i.e., the mean free path of ions between two subsequent collisions is larger than the sheath thickness), the positive ions bombard the cathode with kinetic energy of several 100 eV. If not, the ions scatter within the sheath and the kinetic energy is correspondingly smaller. In any case, practically all kinetic energy gained by ions within the sheath is spent for heating the cathode. A substantial part of the discharge power is therefore spent heating the cathode. This is one of the reasons why the simple calculation of the gas temperature, as shown for the case of atmospheric plasmas in Figure 3.3, is obsolete here. Other reasons will be explained below.

A small part of the kinetic energy of positive ions is spent for kinetic effects such as kicking out electrons and neutral atoms from the cathode surface. These effects are called secondary electron emission and sputtering, respectively. The emitted electrons obviously accelerate within the sheath and gain significant energy. The electrons then thermalize upon elastic collisions with plasma electrons (Mozetič 2011, 174028). This is the main mechanism for introducing discharge power into gaseous plasma. Some positive ions may suffer a charge-exchange collision next to the cathode. These fast neutrals may simply reflect from the cathode since they are not affected by the electric field. The probability for such events is rather low.

Due to the abundance of fast electrons next to the cathode, the plasma glows more extensively next to the sheath, and so the volume of fast electrons is visible with the naked eye. The sheath next to surfaces of materials which are not biased by the DC power supply is thinner than the sheath next to the cathode. The thickness is of the order of several Debye lengths. The heating of such non-biased materials by positively charged ions is thus much smaller than the heating of the cathode, and other heating mechanisms may prevail (Mozetič 2011, 174028). Electrically non-conductive materials such as biological samples cannot be biased, so they are not heated substantially by positive ions, especially when a weakly ionized low-pressure plasma is used. The non-conductive materials exposed to such plasma are kept at so-called floating potential, which is often about −10 V (Primc et al. 2011, 022129).

3.5.4. Heterogeneous surface recombination

Neutral reactive particles, such as metastable molecules or atoms as well as radicals in the ground state, may also lose their energy on the surface of any object facing plasma. Since they do not have an electrical charge, they diffuse freely in the bulk plasma and the sheaths, and finally reach the surface. The interaction between neutral particles and solid material is therefore purely potential, with no kinetic effect. The surface reactions include relaxation of metastables and recombination of radicals to stable molecules, in particular association of neutral atoms in the ground state to parent molecules. All these reactions are exothermic, so they contribute to heating of all surfaces, not only the cathode (Primc et al. 2017, 158-164).

This is another reason why the calculations whose results are shown in Figure 3.3 are obsolete in the case of low-pressure plasma. The probability for such surface reactions depend significantly on properties of materials facing plasma, in particular on the surface composition, morphology, and also the surface temperature. As a general rule, a rougher surface will exhibit more frequent loss of reactive plasma species.

Of particular importance is the loss of atoms by surface association to parent molecules. The effect is often called heterogeneous surface recombination. The probability for such surface recombination depends enormously on the composition of material facing plasma. Although specific to the particular material, a generally observed fact is that this probability is much lower on polymers and glasses than it is on metals. For example, the probability for recombination of oxygen, nitrogen or hydrogen atoms in the ground state on the smooth surface of a polymer was found to be close to 10^{-3} (Zaplotnik et al. 2018, 1800021). On the other hand, there are transition metals like cobalt, nickel, iron, and chromium, where the probability for recombination is about 0.2 for very smooth surfaces, and approaches 1 for extremely rough (on the sub micrometer scale) surfaces. Smooth stainless steel exhibits a recombination probability close to 0.1. The extremely high loss of atoms on nanostructured surfaces has been explained by atoms becoming trapped in gaps with a thickness below the mean free path (Mozetič et al. 2015, 207-213). Such atoms suffer numerous collisions with the surface before they can escape, so the macroscopic loss probability is large.

Glasses are usually smooth. Furthermore, they are chemically very stable, so they do not represent a significant drain of neutral reactive species. In this sense, quartz glass is one of the best materials. The coefficient for heterogeneous surface recombination on a smooth quartz surface is below 10^{-3} for oxygen, nitrogen or hydrogen atoms (Krongelb and Strandberg 1959, 1196-1210; Wood and Wise 1962, 1049-1053). The concentration of neutral reactive species in vacuum chambers made from quartz glass is therefore much larger than in metallic chambers – not due to more intense production, but due to suppressed loss of reactive plasma species on the reactor surfaces. This is the major reason why plasma reactors useful for chemical modification of different materials are made from glass. Alternatively, the reactors may be made from metals, with inner walls coated with quartz glass or a similar material. In many cases, the dissociation fraction in a low-pressure plasma sustained in quartz glass tubes exceeds 50 %, although the ionization fraction remains low, typically about 10^{-6} (Mozetič 2011, 174028). The concentration of neutral reactive atoms in such plasma is therefore almost a million times larger than the concentration of ions. Such plasma is extremely useful for treatment of organic materials which do not withstand heating to elevated temperatures.

3.5.5. DC discharges

A variety of electrical discharges are suitable for igniting and sustaining gaseous plasma in vacuum chambers. The simplest option is a DC discharge. There are two interior electrodes inside a vacuum chamber, of which one is usually powered and the other one grounded. The grounded electrode may simply be the metallic housing of the vacuum chamber (Conrads and Schmidt 2000, 441-454). The DC voltage supply should be equipped with a current limitation, otherwise the plasma may become conductive enough to enable a shortcut between the electrodes, and thereby, transformation of the discharge into a low-impedance electrical arc.

The DC discharge is not useful for sustaining plasma in dielectric tubes due to the requirement for electrodes to be in physical contact with gaseous plasma. As mentioned above, a metallic electrode represents a huge sink of reactive gaseous species due to the high probability of heterogeneous surface recombination. Furthermore, the cathode should be of rather large surface, otherwise the ion flux may be large enough to cause a transition of the discharge regime from a glow discharge to an electric arc. Arcing is detrimental since the discharge power is concentrated on one cathode spot, causing the cathode material to evaporate from the surface and deposit anywhere on the surface of the dielectric tube. A low-frequency AC discharge behaves similarly to the DC discharge, with the exception that the electrodes alternatingly act as the cathode and the anode.

3.5.6. Radio-frequency discharges

The use of high-frequency discharges, such as radio-frequency (RF) or even microwave (MW) discharges, overcomes the drawbacks of the DC discharge. Namely, in these discharges, the electrodes are placed outside the vacuum chamber and sustain a plasma inside the chamber. The chamber should be made from a material of low electrical conductivity to benefit from the electrode placement outside the vacuum chamber (Conrads and Schmidt 2000, 441-454; Mozetič et al. 2015, 015026). Quartz glass is a standard, although soft glasses are often applied as well. Ceramics would perform equally well, but potential drawbacks are porosity as well as opacity of ceramic materials. The latter renders it difficult to monitor plasma conditions, for example by optical emission spectroscopy, which requires transparency of the plasma reactor.

The basic configuration of a vacuum chamber useful for sustaining a plasma with an electrodeless RF discharge is shown in Figure 3.9. The discharge tube is usually made of glass. There is a ring electrode at the central part of the vacuum chamber. The ring electrode is connected to a high frequency power generator, specifically an RF generator, via a matching network (not shown). The matching network serves for adjustment of the impedances. Namely, unlike the DC discharge, a significant fraction of the power available from the RF generator may simply be reflected back to the generator (Butterbaugh et al. 1990, 916-923). Not only does this effect worsen the energy efficiency, it may also be harmful for the generator.

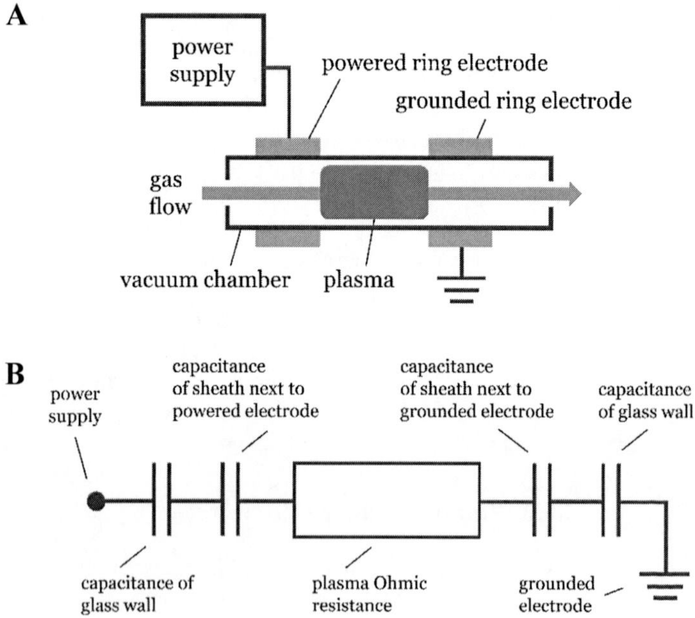

Figure 3.9: An electrodeless RF plasma configuration. (A) Schematic representation. (B) The equivalent electric circuit.

The voltage between the powered and grounded electrodes should be larger than what is required by the Paschen curves at a given pressure (Figure 3.8). In case pressure is too high, or the distance between the electrodes too large, it is not possible to ignite the plasma inside the dielectric tube. One method to overcome this problem is the introduction of a movable electrode inside the dielectric tube and ignition of the discharge between the powered electrode and the electrode inside the dielectric tube. Once the discharge is ignited, the voltage needed to sustain the plasma between the electrode and the grounded electrodes is lower than the voltage in Figure 3.8.

The gaseous plasma in the configuration of Figure 3.9 expands between the powered electrode and one or both of the grounded electrodes. Electrode sheaths are observed at both the powered and grounded electrodes. Although the RF generator provides sinusoidal voltage with equal positive and negative voltages, the glass surfaces next to the electrodes assume a large negative potential of voltage, similarly as in a DC discharge. This effect is called self-biasing and is unavoidable in RF discharges (Conrads and Schmidt 2000, 441-454). The voltage in the sheath oscillates with the same frequency as the RF generator, but is shifted to a large negative self-biasing, so the wall potential becomes positive against the plasma potential only for a short fraction of period time. Such a short time of positive wall potential is sufficient to attract enough electrons to keep the time-averaged fluxes of electrons and positive ions equal. The positive ions bombard the walls next to the electrodes with the energy gained in the DC biased potential drop between the gaseous plasma and the walls next to the electrodes. The rest of the dielectric tube, i.e., between the electrodes, is kept at floating potential, often about −10 V against plasma (Primc et al. 2011, 022129). The electric field penetrates into the gaseous plasma, but attenuates at a short distance from the electrode. This distance obviously decreases with increasing density of electrons in the gaseous plasma.

3.5.6.1 Capacitively coupled RF discharges

The equivalent electrical circuit for the configuration in Figure 3.9 (A) is shown in Figure 3.9 (B). The sheath between the bulk plasma next to the walls has a capacitive character of impedance, so it is represented by a capacitor. This is the reason that the electrodeless configuration presented in Figure 3.9 is called a capacitively coupled (CC) RF discharge. The plasma itself has an Ohmic character, so it is represented by a resistor connected in series with both capacitors. The current carriers within the bulk plasma are free electrons, which have a high mobility due to their low mass. The impedance of bulk plasma is typically smaller than the impedance of both capacitors. The plasma resistance increases linearly with increasing distance between the electrodes. At a large distance or elevated pressure, the resistance may become larger than the impedance of the capacitors (Conrads and Schmidt 2000, 441-454; Lieberman and Lichtenberg 2005).

The configuration in Figure 3.9 is symmetrical, i.e., the powered electrode has the same surface area as the grounded electrodes. Another possibility is an asymmetrical configuration. In that case, the powered electrode is much smaller than the grounded electrodes, and the sheath of large self-bias occurs practically only next to the powered electrode.

3.5.6.2 Inductively coupled RF discharges

If the powered electrode in Figure 3.9 is replaced with a coil wrapped around the outside of the reactor chamber, such a configuration is called an inductively coupled (IC) discharge. One connection of the coil is powered, and the other may or may not be grounded. An electric field is induced by a time-varying magnetic field formed by the RF current passing through the conductive coil (Cvelbar and Mozetič 2007, 24-33). If the other side of the coil is not grounded, the coil represents an antenna. In that case, standing waves are formed along the coil. The powered side of the coil has a similar effect as the powered electrode in Figure 3.9: a sheath of high self-bias is established next to the glass wall backed by the powered side of the coil (Conrads and Schmidt 2000, 441-454).

The discharge properties of the IC discharge are similar to those in Figure 3.9 and can be represented by the equivalent circuit shown in Figure 3.9 (B). This configuration is often called an IC plasma in the E-mode.

3.5.6.3 E-mode and H-mode

Inductively coupled RF plasma may assume one of two modes. The so-called E-mode is characterized by lower electron densities and lower luminosity, while the H-mode is characterized by higher electron densities and higher luminosity (El-Fayoumi et al. 1998, 3082-3094). In this denomination, "E" stands for the electric field and "H" for the magnetic field.

The IC plasma E-mode is described in subsection 3.5.6.1. Its electron density is similar to the electron density in a capacitive RF discharge or in the DC discharge. Typical values are roughly around 10^{16} m^{-3} (Zaplotnik et al. 2011, 55001). The density of neutral atoms is much larger, since the material facing plasma is glass. The typical atom density in a quartz tube is of the order of 10^{21} m^{-3} for the case of hydrogen or oxygen plasma (Zaplotnik et al. 2018, 1800021), and 10^{20} for nitrogen plasma (Kolar and Primc 2016). The difference arises from the fact that the dissociation energy of a nitrogen molecule is twice as much as that of oxygen or hydrogen.

If we increase the available RF power, the electron density in the E-mode increases as well. At a specific discharge power, as the electron density reaches a high enough value, the electrical conductivity of the plasma becomes sufficiently large for the plasma to represent a conductive medium. As this occurs, the plasma can be represented as a one-turn coil, and a transformer effect occurs. This effect is called transition of the IC plasma from the E- to the H-mode. Upon increasing the power absorbed by the plasma, the transition occurs abruptly at a given power, which depends on the discharge geometry and the gas pressure (Kregar et al. 2011, 2502-2503). The abrupt change in the discharge regime is observable with the naked eye, as the plasma luminosity jumps for a few orders of magnitude. The density of electrons in the H mode is typically above 10^{18} m^{-3} (El-Fayoumi et al. 1998, 3082-3094). The atom density, however, does not change much, since the 50 % dissociation degree has already been achieved in the E-mode. Plasma in the H-mode is a significant source of radiation, in particular in the UV and VUV range of wavelengths. As much as 10 % of discharge power is turned to radiation in the H-mode. In the E-mode, the radiation usually accounts for less than 1 % of the discharge power.

The transition from E- to H-mode is abrupt upon increasing the power at the RF generator, and so is the transition from the H- to E-mode when the RF power decreases. However, the transition from high-density plasma in the H-mode to the low-density plasma in the E-mode does not occur at the same power as the transition from the E- to the H-mode. Instead, a hysteresis is observed. The transition from E- to H-mode always occurs at a higher power of the RF generator that the opposite transition (Zaplotnik et al. 2011, 55001). The transitions are more pronounced at elevated pressures. The IC plasma in the H-mode is not very useful for treatment of the organic material, since the thermal load is usually too large.

It should be added that in an IC RF plasma system, an electrostatic field is present due to potential differences between the induction coil and the ground surfaces of the system (Bol'shakov et al. 2004, 823-832). At lower input powers, in the E-mode, this electrostatic field is larger than the induced field, and thus, discharges primarily form as a consequence of capacitive coupling. In the H-mode, the induced electric field is larger, and is the primary factor sustaining the discharge. At the same time, the electrostatic field, and thereby some capacitive coupling, remains present as an additional aspect of plasma, known as stray capacitance (El-Fayoumi et al. 1998, 3082-3094).

3.5.6.4 Glow and afterglow

Gaseous plasma glows due to relaxation of electronically excited atoms or molecules. As soon as the electric field is switched off, the electron temperature drops almost instantly (over a time scale well below a microsecond). As the electron temperature drops, the excitation of the radiative states becomes inefficient, so the plasma stops glowing. For a short time, often of the order of a microsecond, the charged particles still persist, but the electron temperature is too low to enable excitation to radiative states. This short period after turning off the plasma generator is called "early afterglow" (Stevefelt and Robben 1972, 1502-1515). The electrons and ions keep diffusing and eventually reach the reactor walls where they neutralize. As a result, only neutral reactive particles remain for the next millisecond or so. This step in the discharge evolution is called "late afterglow" (Nafarizal et al. 2008, 035206). The concentration of neutral reactive particles decreases with increasing time after turning off the discharge, and eventually, the gas assumes thermal equilibrium. The time needed for an almost full thermalization obviously depends on the diffusion of neutral reactive species and the surface loss rates. As explained above, many materials exhibit low recombination coefficients, so the life-time of the late afterglow in a plasma chamber may be well over a millisecond, often in the range of 0.1 s. Both early and late afterglows are therefore rich in neutral reactive species, but poor in energetic charged particles and radiation. The afterglows are therefore useful when one wants to treat materials with neutral radicals only. The obstacle is the rather low life-time of the afterglow.

This obstacle is overcome by using flowing afterglows (Vesel et al. 2018, 094005). In such a case, the gas passes the volume of electrical discharge, where it is transformed to plasma, and then continues its way towards a vacuum pump. The distance from the discharge region in flowing afterglow has the same role as time in temporal afterglow: if the gas drift velocity is v, the time of afterglow is $t = s/v$, where s is the distance from the discharge. For example, when the gas drift velocity is 10 m/s (a typical value), the distance from the discharge zone of 1 m corresponds to the time of 0.1 s after turning off the discharge. As mentioned earlier, the life-time of the late afterglow is of the order of 0.1 s, so even 1 m away from the discharge zone, the concentration of neutral reactive species is significant, perhaps an order of magnitude lower than in the glowing plasma itself. Flowing afterglows are therefore very useful for continuous treatment of sensitive objects, such as biological matter, with neutral reactive plasma species, without the need for pulsing the discharge.

Obviously, a spatial evolution of the reactive species density exists along a continuously pumped plasma reactor. On one side, working gas is introduced at a constant rate, and on the other, the vacuum chamber is pumped continuously. Such vacuum chambers are often long tubes made from dielectrics – as explained above, glass is the material of choice. Using such a configuration, one can choose between treating samples with glowing plasma, early afterglow, or late afterglow, simply by placing the object in the appropriate segment of the vacuum tube (Gorjanc and Mozetič

2014; Kutasi et al. 2014). Ions and excited metastable atoms will typically undergo recombination outside of the immediate discharge area, but neutral excited atoms and molecules will be present in the afterglow as well (Mozetič et al. 2015, 015026).

3.5.7 Low-pressure oxygen plasma chemistry

Plasma discharges in different gaseous media produce reactive species based on the chemistry of the gas. Low-pressure oxygen plasma is commonly used in experimental practice, and has been well characterized regarding the reactive species densities and behavior (Cvelbar and Mozetič 2007, 24-33). A wide assortment of reactive oxygen species (ROS), deriving from the ground-state oxygen molecule O_2, is found in low-pressure oxygen plasma. These include:

- O_2, the oxygen molecule in the ground state, also denoted $O_2(X^3\Sigma_g^-)$;
- O_2^+ and O_2^-, its molecular ions;
- $O_2(a^1\Delta_g)$ and $O_2(b^1\Sigma_g^+)$, the metastable, excited singlet states of O_2;
- O, the biradical neutral oxygen atom in the ground state, also denoted $O(^3P)$;
- $O(^1D)$ and $O(^1S)$, the metastable singlet states of the oxygen atom;
- O^+ and O^-, ions of the oxygen atom;
- O_3, ozone;
- O_3^+ and O_3^-, its ions (Toneli et al. 2015, 325202).

In a plasma, the electrons, gas molecules, and the various reactive species interact with one another, resulting in a complex chemistry. Several types of reactions occur (Cvelbar and Mozetič 2007, 24-33), and some of the most important are listed below. In a dissociation reaction, a molecule dissociates upon impact with an electron (e). This is an important process in low-pressure discharges, as it produces energetic neutral particles which then react with the sample surface. In case of molecular oxygen (Lieberman and Lichtenberg 2005), the reaction proceeds as

$$e + O_2 \rightarrow 2O + e. \tag{3.2}$$

On the other hand, ionization results in the formation of an ion, either molecular (electron–molecule ionization) or atomic (electron–atom ionization), via interaction of an incident electron with a valence electron of a neutral atom or molecule (Fridman 2008):

$$e + O_2 \rightarrow O_2^+ + 2e, \tag{3.3}$$

$$e + O \rightarrow O^+ + 2e. \tag{3.4}$$

At lower electron energy, ionization is less likely to occur than dissociation. However, ionization probability increases with increased electron energy, and eventually predominates at electron temperatures above 100 eV, which are unlike for gaseous plasma suitable for treatment of organic matter (Cvelbar and Mozetič 2007, 24-33).

The processes of reactive species formation and loss run in parallel. The process by which reactive species combine to form parent molecules is called recombination. The most important molecular ion dissociative recombination reaction in the oxygen plasma gas phase is

$$e + O_2^+ \rightarrow 2O. \tag{3.5}$$

The loss of neutral atoms in the gas phase is predominantly in three-body collisions, so its probability at low pressures is rather low. Nonetheless, surface recombination also occurs at the surfaces of either the reaction chamber or any materials exposed to the plasma, and can be an important reactive species loss mechanism for ions as well as neutral oxygen atoms in low-pressure systems (Zaplotnik et al. 2018, 1800021; Cvelbar and Mozetič 2007, 24-33). However, the probability for the surface reaction

$$O + O \rightarrow O_2 \tag{3.6}$$

depends on the recombination coefficient of the material, which is rather low for either quartz or borosilicate glass, so neutral atom losses are not substantial on the walls of glass reactors (Kregar et al. 2011, 2502-2503). As explained in subsection 3.5.6.4, a significant amount of the oxygen atoms is pumped away from the system rather than lost by recombination, providing that the surface loss coefficient is low. Further, as explained in subsection 3.4.1, gas phase loss of atoms is marginal at pressures below several mbar.

The life-time of both molecular metastables, $O_2(a^1\Delta_g)$ and $O_2(b^1\Sigma_g^+)$, is enormous: 45 minutes and 10 seconds, respectively (Mozetič 2011, 174028). Further, it is presumed that their relaxation on glass surfaces is negligible. The molecular metastables therefore persist in late afterglows. Unfortunately, little literature is available on interaction of these metastables with organic molecules. The same applies for atomic metastables. The radiative life-time of the first atomic metastable state, $O(^1D)$, is about 110 s (Young et al. 1968,

4758-4768), so these species remain in the gas phase as long as they are not relaxed on surfaces. The radiative life-time of the $O(^1S)$ state is about 200 μs, but the excitation energy of over 4 eV is large enough to relax on the surfaces extensively, so these species should only be taken into account in the glowing plasma or early afterglow (Vialle et al. 1991, 301-308). Ozone is stable at room temperature in equilibrium gas, but in non-equilibrium conditions, it may interact with metastables and/or atoms in the ground state to form O_2 molecules. It should be stressed here that the science on the kinetics of metastables in plasma reactors, and particularly their interaction with surfaces, is still in its infancy.

On the other hand, the behavior of neutral oxygen atoms in the ground state in plasma and afterglow conditions has been elaborated by numerous authors. It is generally accepted that the O-atoms persist in both early and late afterglow and are regarded the major reactants in weakly ionized oxygen plasma suitable for treatment of organic materials.

4. Plasma Reactors and Clove Treatment

4.1 Plasma reactors

During the course of this book, two example plasma systems are selected for the plasma treatment of garlic cloves. Both use a low pressure, inductively coupled, radio frequency oxygen plasma to treat the garlic cloves. The main difference between the systems is their size; one is a pilot-scale system and the other an industrial-scale system.

The components of both systems are as follows. The reactor is a borosilicate discharge tube, either 800 mm in length and 36 mm inner diameter (pilot-scale system) or 2000 mm in length and 200 mm in diameter (industrial-scale system). To sustain low pressure, the reactor is continuously pumped using a vacuum pump of an ultimate pressure well below 1 Pa. Oxygen is introduced into the reactor by a flow controller, and pressure is monitored using an absolute pressure gauge. A copper coil, either water- or air-cooled, is wrapped around the reactor tube. A radio frequency power generator is matched to the coil via a matching network. A schematic representation of the systems is shown in Figure 4.1, and a photograph of the pilot-scale system is shown in Figure 4.2.

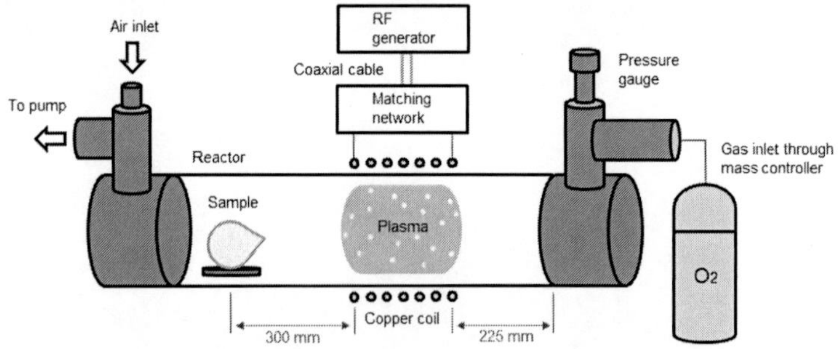

Figure 4.1: The pilot-scale plasma system. A schematic representation, showing an H-mode plasma and a sample in the afterglow treatment position.

Figure 4.2: Photograph of the pilot-scale plasma system. Side view during H-mode afterglow treatment of a garlic clove, with indicated locations of the pump (1), air inlet valve (2), pressure gauge (3), and gas inlet (4).

4.2 Pilot-scale plasma treatment

For treatment in the pilot-scale plasma system, the sample clove is placed onto a microscope slide and positioned inside the glass reactor tube. The E-mode plasma is fairly uniformly distributed along the discharge tube, and treatment in E-mode glow plasma implies a clove position at the center of the copper coil. On the other hand, H-mode afterglow plasma implies that the clove is positioned downstream from the copper coil, in the direction of the vacuum pump, as illustrated in Figure 4.1. While H-mode plasma is mostly spatially enclosed within the coil, stray capacitive coupling remains present along the discharge tube. Therefore, such a treatment is clearly not a "true" afterglow, which would require only the presence of long-lived plasma reactive species. Nonetheless, for the sake of simplicity, this treatment will be addressed simply as "afterglow" throughout this book. It should further be stressed that the capacitive component of coupling is marginal when the plasma is in H-mode, so the radiation is hardly visible outside the coil, indicating poor density of charged particles and thus practically no kinetic interaction, as compared to the potential interaction between the neutral oxygen atoms in the ground state and the organic matter.

As an example, let us consider a set of discharge parameters applied in experiments. For pressures ranging from 30 to 70 Pa, and nominal power (P_N) at the generator ranging from 100 to 300 W, the remainder of the parameters is collected in Table 4.1.

Table 4.1: Discharge parameters and sample positions. Clove treatment in the pilot-scale plasma system. Sample position: glow (G) or afterglow (AG). p is the pressure inside the tube during plasma treatment. PN, PFW, PRF, and PEFF are the nominal, forward, reflected and effective powers of the particular treatment, respectively.

p [Pa]	P_N [W]	P_{FW} [W]	P_{RF} [W]	P_{EFF} [W]	mode	position
30/50/70	100	100	75	25	E-mode	G
30/50/70	200	200	150	50	E-mode	G
30/50/70	300	275	200	75	E-mode	G
30	300	300	30	270	H-mode	AG
50	200	200	15	185	H-mode	AG
50	300	300	15	285	H-mode	AG

4.3 Industrial-scale plasma treatment

In the industrial-scale plasma system, the plasma luminosity and reactive species density is highest within the coil, but capacitive coupling is present throughout the tube. Cloves are placed on a glass plate and positioned inside the reactor tube. For treatment in glow plasma, the plate is placed at the center of the copper coil. Several cloves may be treated simultaneously, but care should be taken to ensure sufficiently low pressures in a reasonable amount of time. The garlic samples are always kept at floating potential; thus the kinetic effects are marginal.

4.4 Control treatment

To confirm that any recorded changes or improvements to the garlic clove are due to plasma treatment, not other factors, control treatments should always be included. In the case of radio frequency, low-pressure oxygen plasma, these may include: (1) untreated control; (2) vacuum control, which exposed cloves to low pressure and oxygen gas flow, without turning on the plasma; and (3) electromagnetic (EM) field control, which exposed cloves to the EM field while plasma was not ignited.

5. Reactive Species Identification

5.1 Method principle

To understand and directly compare the properties of any plasma used in experiments, it is highly recommended to characterize the discharge. The following chapters deal with the properties of the plasma discharge itself, as well as the discharge during treatment of garlic clove samples. We consider two plasma characterization techniques: optical emission spectroscopy (OES) as well as catalytic probe measurements.

OES provides information on the reactive particles in the plasma. The particles emit electromagnetic radiation at wavelengths characteristic of a particular element or particle. To obtain a OES spectrum, we record light intensity across a range of wavelengths. The intensity of the signal at a wavelength specific for a chemical species is called an emission line. Signal intensity is measured in arbitrary units and provides some information about the relative quantity of that chemical species in the discharge, but absolute values cannot be directly compared. The spectra shown in this chapter are recorded in the range from 200 to 1100 nm, using an integration time of 1000 ms for the case of measuring radiation in the E-mode.

5.2 E-mode glow plasma

The typical configuration of a low-pressure plasma system during may be called a "flow configuration", as the system is continuously pumped, gas is continuously introduced, and the steady flow of gas maintains a steady working pressure. The OES spectra presented here are recorded for discharges in the flow configuration. First, let us consider OES spectra of the pilot-scale E-mode plasma, as shown in Figure 5.1.

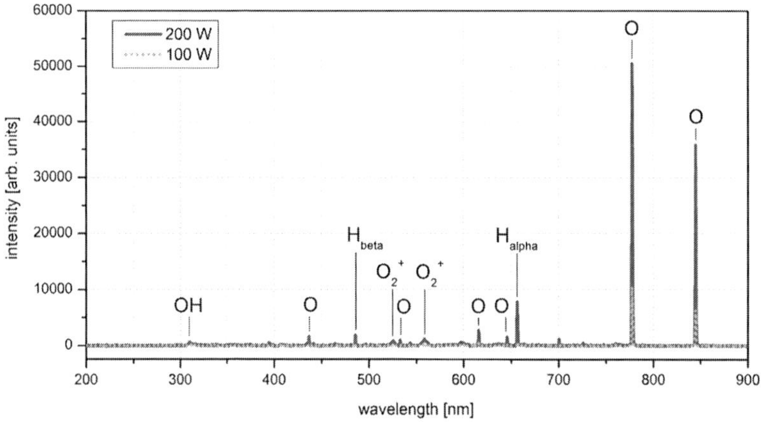

Figure 5.1: OES spectra, pilot-scale E-mode oxygen plasma. P_N = 100 –200 W, p = 50 Pa, flow configuration. Spectra recorded approximately 60 s after turning on the discharge.

The main species detected by OES in E-mode plasma alone is the neutral oxygen atom (O). At the discharge parameters of $P_N = 100$ W and $p = 50$ Pa, oxygen atomic emission lines at around 777.4 and 844.6 nm are prominent (Zaplotnik et al. 2016, 144), along with an additional weak O line at 616 nm. After increasing P_N to 200 W, the O lines at 616, 777.4, and 844.6 nm are about 5 – 7 times more intense than at $P_N = 100$ W. Several additional weak O lines appear at 437, 486, 533, 544, 645 and 926 nm (Musil et al. 1993, 533-540; Kregar et al. 2012, 25-30).

The prevalence of oxygen atoms as the primary reactive species is typical for low-pressure RF oxygen plasma (Kregar et al. 2011, 2502-2503), and the density of positive ions in such plasma is several orders of magnitude smaller than that of neutral atoms. The reasons for this general rule have been elaborated elsewhere (Vesel and Mozetič 2017), but for the sake of completeness, let us mention the key reason relevant for plasma sustained in glass tubes by electrodeless RF discharges: while the probability for surface neutralization of positively charged ions is almost 100 % regardless of the surface properties, the probability for the surface loss of atoms depends enormously on the type of material facing plasma. For many glasses, the probability for the loss of atoms by heterogeneous surface recombination is below 10^{-3} (Mozetič et al. 2015, 015026). Furthermore, the dissociation energy of oxygen molecule is about 5 eV, while the ionization energy is close to 12 eV. There are a couple of long-living metastable states of neutral oxygen molecule at potential energy of about 1 and 2 eV, so stepwise dissociation is likely to occur. For these reasons, the oxygen atom density is as large as 8.5×10^{21} m^{-3} in the pilot-scale reactor, as shown in Figure 6.2.

As seen by the presence of oxygen molecular ion (O_2^+) and atomic ion (O^+) lines, ionization still occurs, but is not nearly as prominent (Cvelbar and Mozetič 2007, 24-33). At P_N = 100 W, a weak line of O_2^+ is present at 558.7 nm. Even after increase to P_N = 200 W, the O_2^+ lines, e.g. at 525.5 and 558.7 nm, are still very weak. Even weaker are the O^+ lines seen at 441.5 and 464.9 nm (Zaplotnik et al. 2016, 144). It is also not expected that these ions would have high kinetic energy gained in the sheaths of the RF discharge. The relevant positive ions in oxygen plasmas are O_2^+ and O^+, but dissociative ionization is energetically less favorable than ionization of molecules (Mozetič 2010, 165-171), which explains why the O_2^+ lines are relatively more intense than the O^+ lines.

In addition to oxygen, the hydrogen Balmer line H_α is detectable at around 656 nm (Kramida et al. 2018), and is about 12 times more intense after increasing the P_N from 100 to 200 W. The OH band also appears weakly at around 309 nm (Zaplotnik et al. 2016, 144). The presence of the OH band and hydrogen H_α line is the consequence of residual water vapor present in the vacuum system as an impurity (Cvelbar et al. 2007, 224-227). See discussion in subsection 3.5.1.

Within the given reactor, a higher nominal power such as P_N = 300 W results in a transition into H-mode plasma at both 30 and 50 Pa whenever a discharge without a garlic sample is turned on.

5.3 Garlic clove treatment

During treatment of peeled or unpeeled cloves in the E-mode glow plasma (Figure 5.2), the signal pattern is generally the same as when no garlic is present (Figure 5.1). The main difference is in the considerable changes in the intensity of the individual spectral lines. The oxygen atomic lines are decreased to about 50 % intensity if a garlic clove is present in the plasma. Simultaneously, the H_α and OH lines increase by a factor of about 2 and 4 for unpeeled and peeled cloves, respectively. The reason is desorption of water vapor from the organic matter upon exposure to vacuum conditions.

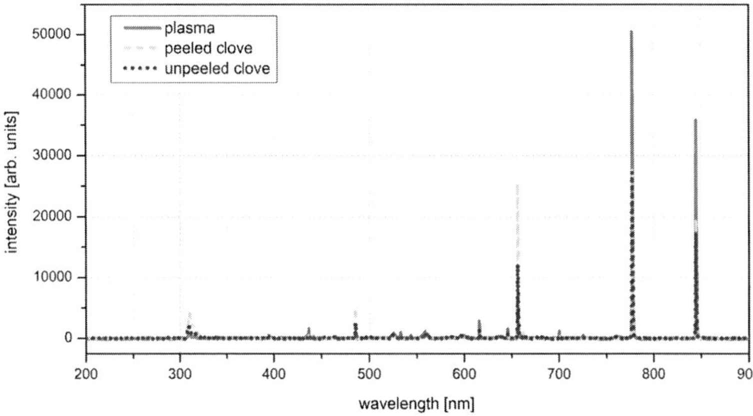

Figure 5.2: OES spectra, pilot-scale E-mode clove treatment. $P_N = 200$ W, $p = 50$ Pa, plasma alone, as well as treatment of a peeled or unpeeled garlic clove in the flow configuration. Spectra recorded approximately 60 s after turning on the discharge.

The changes in spectra line intensity indicate an important change in the plasma chemistry. Decreased signal intensity of oxygen atomic lines suggests a decreased production rate or an increased loss rate (Cvelbar et al. 2007, 224-227), and can be described by several contributing factors. The garlic clove, peeled or unpeeled, facilitates O atom loss due to the reactions occurring at the clove surface. In the process of functionalization, O atoms chemically incorporate into the surface structure (Wheale et al. 1998, 6699-6704). With further oxidation, due to chemical etching, they are lost from the surface in the form of CO or CO_2, via a mechanism which will be further addressed below. In addition, recombination of O atoms to parent molecules may also occur. While to the best of our knowledge, no data is available regarding recombination coefficients for plant surfaces or waxes, recombination coefficients for polymers are typically on the order of 10^{-3} (Zaplotnik et al. 2018, 1800021). This value is rather low, but still 1 – 2 orders of magnitude greater than that of borosilicate glass (Wickramanayaka et al. 1991, 2897-2900), which forms the reactor walls. Even further, the chemical species released from the clove intervene with the formation of reactive species, or react with them in the plasma volume (Clouet and Shi 1992, 1955-1966), thereby decreasing their density.

Simultaneously, the H_α and OH lines increase. This indicates that the water present in the system is no longer a residual impurity, but actively added into the plasma. The source of the water is the clove itself. Two mechanisms are possible. Firstly, if chemical etching occurs at the surface,

OH radicals form by hydrogen abstraction by the oxygen atom, and may then abstract another hydrogen to form water (Clouet and Shi 1992, 1955-1966). Secondly, if cracks or fissions form in the cuticle (Stolárik et al. 2015, 659-676), this may facilitate water release from the bulk of the clove in the low-pressure environment. The dissociation energy of the water molecule is rather low. Therefore, once in the gas phase, water is dissociated to OH and H radicals by the plasma electrons (Gorjanc et al. 2009, 557-567). This is seen in the OES spectra as an increase in the intensity of the corresponding spectral lines.

The OES spectra during plasma treatment of garlic cloves with an increase of generator power (Figure 5.3) or working pressure (Figure 5.4) both show that the increase in H_α and OH line intensity is larger for peeled than unpeeled cloves, especially at P_N = 100 or 200 W. This indicates that more water is present in the system when a peeled clove as opposed to an unpeeled clove is treated. Since both the protective leaf and the peeled clove are covered with a cuticle, no considerable differences in the etching products are expected. However, compared to the clove itself, which contains about 65 % water (Kamenetsky 2007, 123-172), the protective leaf is relatively dry, with the cells of its parenchyma mainly collapsed (Mann 1952, 195-251). Its function is protection of the clove from environmental effects, including water loss. Therefore, under the assumption that etching and cracking of the treated surface takes place, less water is present in the system during the treatment of the unpeeled clove, firstly, due to a lesser quantity of water in the treated surface, and secondly, due to protection of the water-containing clove from the plasma by the protective leaf.

5.4 Effect of generator power

During plasma treatment of a garlic clove, when the nominal generator power increases, the signal of all important emission lines increases as well (Figure 5.3). This includes the oxygen atom and molecular ion as well as the H_α and OH lines. This observation is unsurprising. In the E-mode plasma, at fixed pressure, reactive species density will gradually increase with increased treatment power (Zaplotnik et al. 2011, 55001), explaining the O and O_2^+ signal intensity increase. This is confirmed by catalytic probe measurements of oxygen atom density (Figure 6.2). These reactive species, in turn, accelerate surface modification, resulting in accelerated etching and water release from the clove, which is evident in the increased H_α and OH lines.

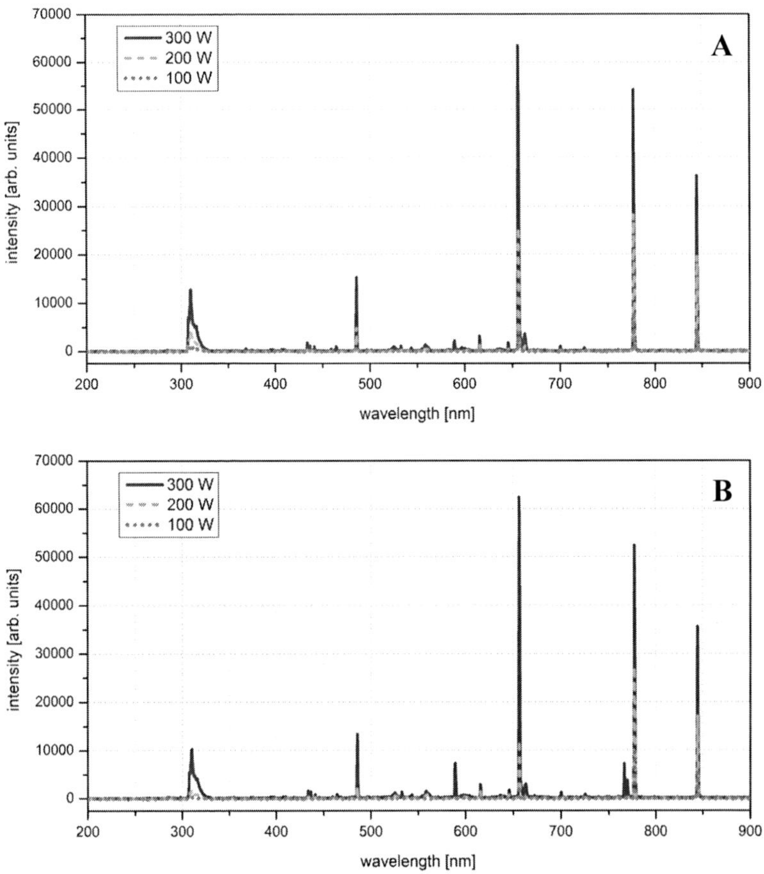

Figure 5.3: OES spectra, pilot-scale E-mode, increasing power. $P_N = 100 - 300$ W, $p = 50$ Pa, flow configuration. Treatment of (A) peeled garlic cloves, (B) unpeeled garlic cloves. Spectra recorded approximately 60 s after turning on the discharge.

Signal increases during treatment of an unpeeled clove at increasing P_N are very similar to the ones seen for peeled clove treatment. One prominent difference is the appearance of spectral lines at approximately 589, 766, and 770 nm. These lines appear due to the presence of sodium (589 nm) and chlorine (766 and 770 nm). Out of all discharge parameters used in our experiments, the 300 W, 50 Pa condition has the highest O atom density. When treating unpeeled cloves, exposed parts of the protective leaf (the thin tip, or any cracks that form due to the heat) may begin to sear due to intense heat generated in the reactions with the plasma species. This causes thermal degradation of the organic compounds in the protective leaf, releasing material into the gaseous phase. This may be the source of elements such as sodium and chlorine detected during this treatment. Here, it should be stated that sodium atoms radiate extensively in low-pressure plasma, so even a small concentration of Na in the gas phase will result in a significant peak in the optical spectrum. In fact, low-pressure sodium lamps are highly efficient electrical light sources and are often used in outdoor lighting, such as street lamps.

5.5 Effect of working pressure

The signal of prominent emission lines during plasma treatment of garlic also increases when the working pressure is decreased at constant generator power (Figure 5.4). However, the difference in O atom line intensities (777.4 and 844.6 nm) at the pressures of 30 and 50 Pa is barely detectable in the case of unpeeled cloves.

If we additionally consider oxygen atom density within the selected plasma parameters range, (Figure 6.2), it either does not change ($P_N = 100$ W) or decreases ($P_N = 200$ W) with increased working pressure. At low pressures, the mean free path that particles can travel is larger, so the electron temperature is increased. Consequently, ionization probability is also higher, and thus the electron density is increased (Cvelbar et al. 2007, 224-227). This results in a more intense excitation of reactive plasma species, detected as an increase in the OES signal intensity.

The accelerated etching allows for more rapid water loss from the clove, resulting in an increased OH emission line signal, which eventually reaches a plateau when an equilibrium is established between the water release from the clove and the pumping speed.

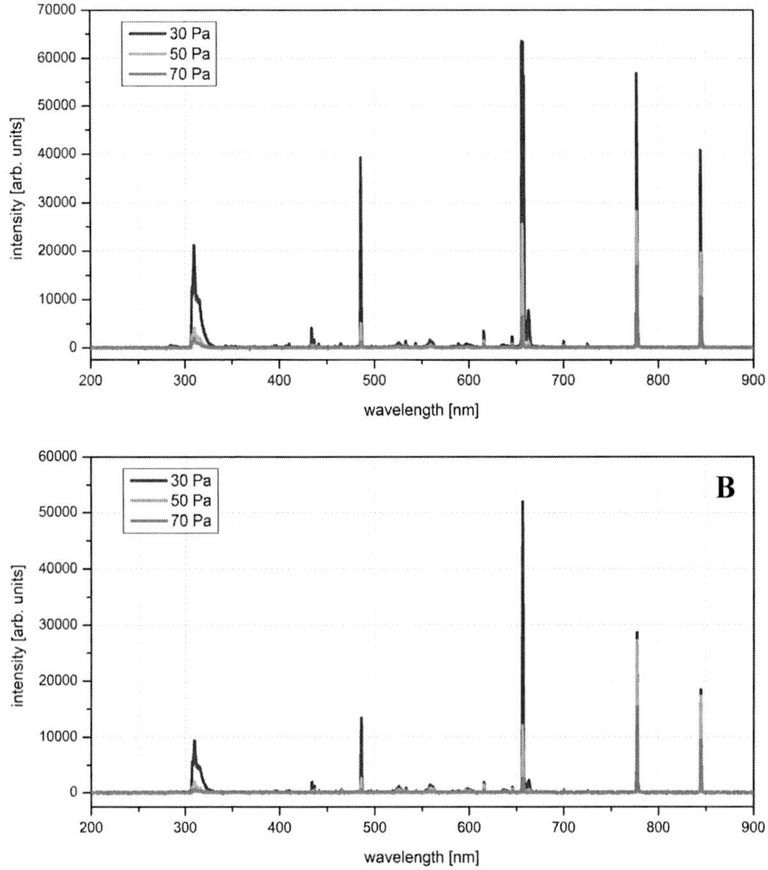

Figure 5.4: OES spectra, pilot-scale E-mode, increasing working pressure. $P_N = 200$ W, p = 30 – 70 Pa, flow configuration. Treatment of (A) peeled garlic cloves, (B) unpeeled garlic cloves. Spectra recorded approximately 60 s after turning on the discharge.

5.6 No-flow configuration

To gain further insight into the dynamics of reactive species in the plasma discharge, we may also consider a so-called "no-flow configuration". To achieve it, the reactor is first evacuated, the pump is turned off, and oxygen is carefully released into the reactor up to the desired working pressure. Turning on the discharge in this configuration allows for the accumulation of plasma species that normally appear in very limited amounts during the course of the experiment, in particular during clove treatment. This makes the signal intensity changes with time significantly more dramatic than in the flow configuration. Thereby, the no-flow configuration facilitates the detection of these species by OES. To better present these dynamic changes, the data is plotted in Figure 5.5 through Figure 5.7 as time courses of signal intensities at selected wavelengths, rather than "static" spectra at a specific time point.

Time courses of selected signal intensities during plasma treatment of peeled cloves in the no-flow configuration are shown below. In Figure 5.5, it is apparent that H_α and OH dominate the spectrum. The remaining signals, including the O lines, rise briefly after turning on the discharge, but subside within about 20 s, when the pressure rises to about 75 – 100 Pa. The H_α and OH lines then steadily decrease while pressure increases and plasma luminosity falls. At about 200 s, the discharge can no longer be sustained due to the pressure rise, and extinguishes by itself.

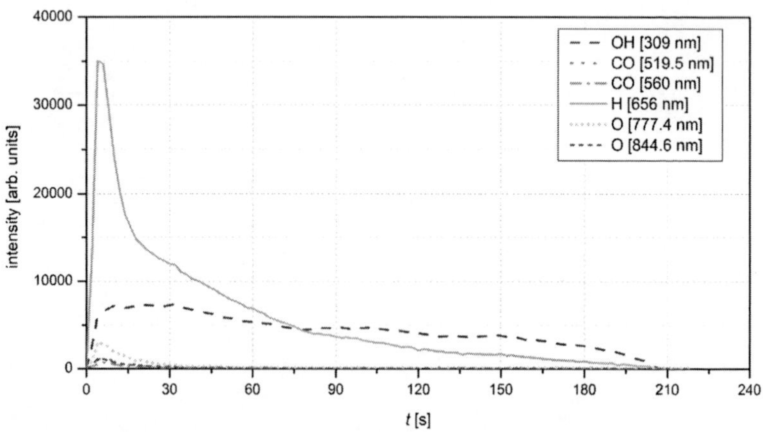

Figure 5.5: OES signal intensity, no-flow. Pilot-scale E-mode plasma, peeled clove, $P_N = 200$ W, $p = 50$ Pa.

Figure 5.6 compares the temporal behavior of selected spectral lines of treatments with P_N = 200 W and P_N = 300 W. Interestingly, while the behavior and intensity of the OH signal is nearly identical, the remaining signals (H_α, O, and CO) are all notably more intense in the P_N = 300 W treatment. In particular, the initial H_α peak is nearly twice as intense. However, the time course of the H_α line intensity from 20 s onward is also comparable between the P_N = 200 and P_N = 300 W treatments.

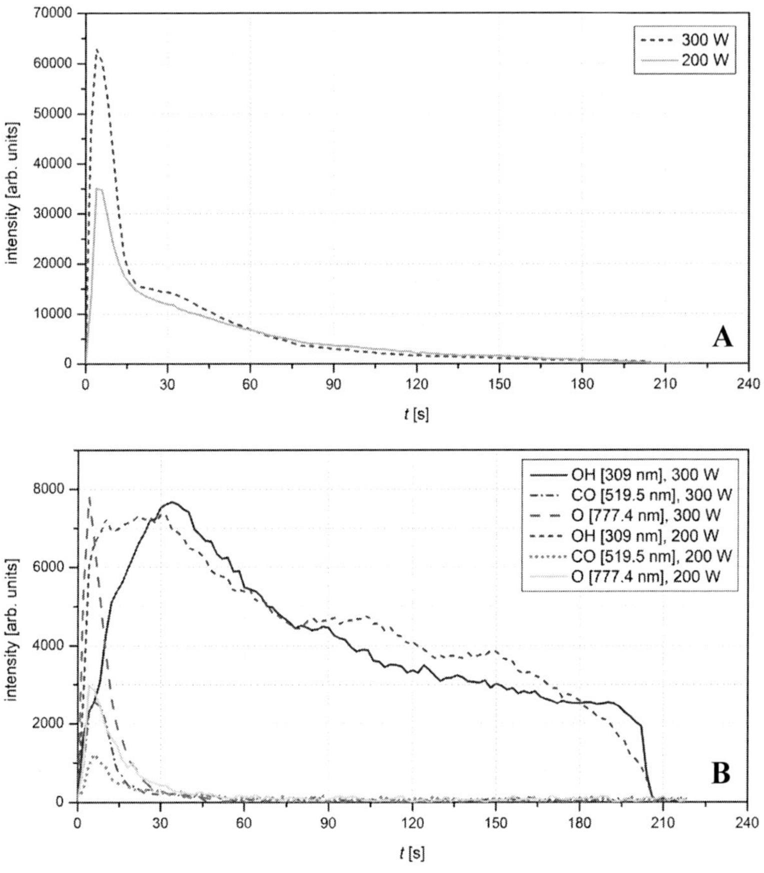

Figure 5.6: OES signal intensity, no-flow, increasing P_N. Pilot-scale E-mode plasma, peeled clove, P_N = 200 – 300 W, p = 50 Pa. (A) H_α; (B) other signals.

We know with certainty that surface etching takes place during plasma treatment of garlic cloves, as confirmed by SEM images of the surface (Figure 9.1 – Figure 9.5) and further explained below. During this process, the organic material is gradually degraded, and carbon is released into the system in the form of volatile compounds, in particular, CO and CO_2 (Vesel and Mozetič 2001, 373-377). When organic material is oxidized in this way, emission lines of CO are typically seen in OES spectra, commonly including CO Angstrom bands at around 519.5 and 560 nm (Vesel et al. 2007, 3689). In the flow configuration spectra, we have seen that CO emission lines are either not present, or seen very weakly. Because the system is pumped throughout, the CO is steadily removed by the pump, and may remain under the detection limit of the OES. To further confirm the presence of CO, and thereby the occurrence of etching, let us now consider the spectra recorded in the no-flow configuration. The initial presence of the CO lines in these charts confirms the occurrence of etching, but the CO signal intensity is still rather low, and subsides rapidly. This is due to the plasma conditions in the no-flow configuration. Firstly, the reactor is not pumped, but the etching and water loss introduce new molecules into the gaseous phase, resulting in constantly rising pressure in the reactor, mainly due to water accumulation. At the same time, no additional O_2 is introduced into the system. All of the above strongly influence the plasma chemistry. As the plasma conditions and chemistry change due to the lack of pumping and oxygen supply, etching subsides, and the CO signal decreases rapidly.

Comparisons of time courses of selected signal intensities during plasma treatment of peeled and unpeeled cloves in the no-flow configuration are shown in Figure 5.7. During the treatment of unpeeled cloves, the plasma lasts longer than it does for peeled cloves (around 330 s, compared to around 200 s). Further, most spectral lines are more intense for the unpeeled clove, particularly the O and CO signals, which are twice as intense as during the treatment of a peeled clove. A notable exception is the OH band, which is more intense during the treatment of the peeled clove. This indicates more intense etching of the unpeeled clove, but more water presence for the peeled clove. More water released during peeled clove treatment also contributes to faster pressure rise, and therefore a shorter-lasting plasma.

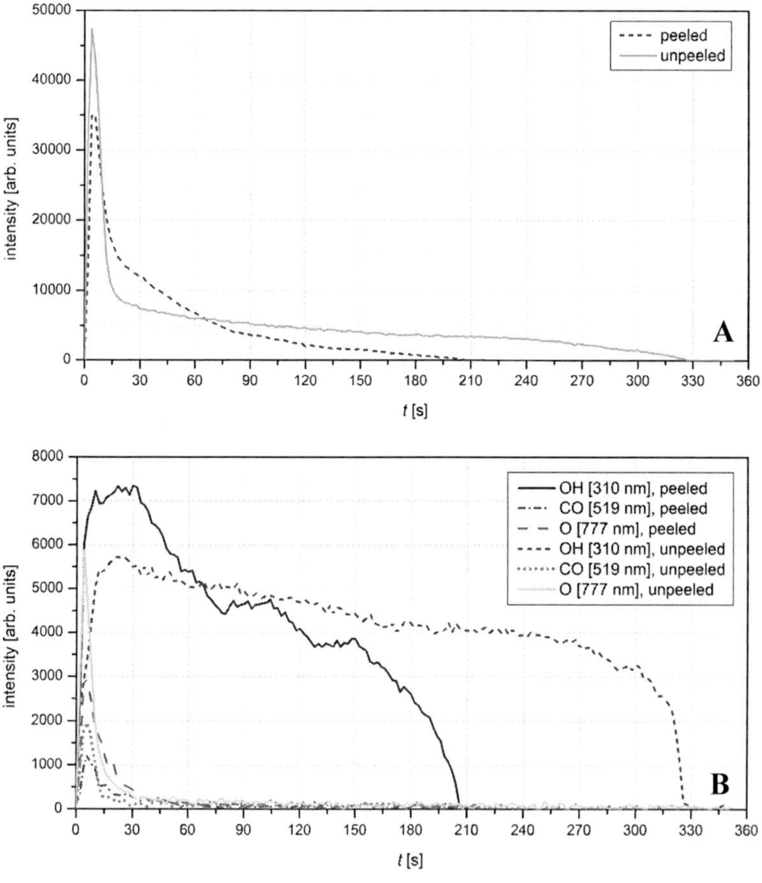

Figure 5.7: OES signal intensity, no-flow, peeled and unpeeled cloves. Pilot-scale E-mode plasma, $P_N = 200$ W, $p = 50$ Pa. (A) Hα; (B) other signals.

6. REACTIVE SPECIES DENSITY

6.1 Catalytic probe measurements

One of the most important characteristics of low-pressure oxygen plasma is the oxygen atom density (n_O) – that is, the density of its most important reactive species. n_O in a discharge can be determined using a catalytic probe. A laser-heated fiber-optics catalytic probe inserted into the plasma reactor tube is seen in Figure 6.1. Such a probe employs a small catalytic tip made from pure cobalt with a surface oxide layer. The tip is mounted onto an optical fiber which connects to an electronic unit consisting of an infrared radiation detector (for measuring tip temperature) and a laser (for heating the tip to elevated temperature). Exothermic reactions – recombinations of reactive plasma species to parent molecules – occur at the probe tip when the discharge is turned on. However, the tip temperature is held constant by controlling the laser power. From the difference in the laser power between the plasma on and off conditions, the flux of reactive plasma species onto the probe tip is calculated. The probe should be placed far enough from the discharge to minimize the effect of ions which could additionally heat the tip surface when impinging on it.

Figure 6.1: Catalytic probe. Shown mounted onto the pilot-scale plasma system for measurements during treatment of cloves in the H-mode afterglow.

6.2 Oxygen atom density in the E-mode plasma

In the pilot-scale E-mode plasma without a garlic sample (Figure 6.2), the n_O increases with increasing generator power. This corresponds well to the OES spectra in Figure 5.1, where the O emission lines intensity increases with increased P_N. At fixed working pressure, n_O is known to increase with increasing generator power (Zaplotnik et al. 2012, 3857-67), and the parameters selected here appear to be in a range of approximately linear increase.

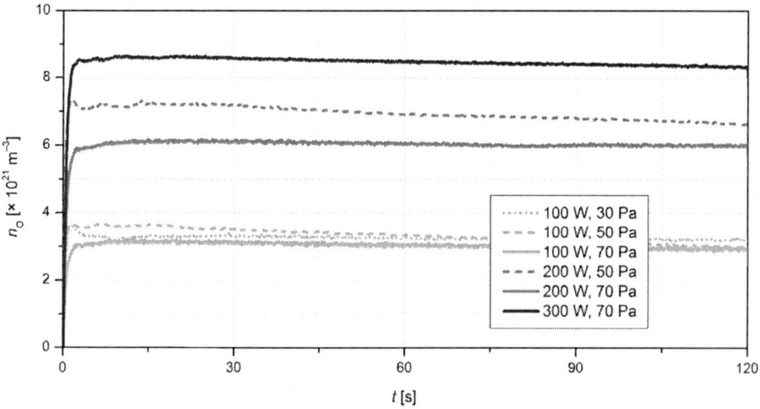

Figure 6.2: n_O, pilot-scale E-mode plasma without garlic sample. $P_N = 100 – 300$ W, $p = 30 – 70$ Pa, $t = 120$ s. The discharge is turned on at $t = 0$ s and turned off at $t = 120$ s.

We know from Table 4.1 that within the selected range of plasma conditions, the P_{EFF} at a constant P_N is also constant, regardless of the working pressure. As we see here, this is reflected in the n_O at $P_N = 100$ W, which is approximately equal ($\sim 3 \times 10^{21}$ m^{-3}) at all three working pressures. At $P_N = 200$ W, the n_O at 50 Pa is slightly higher than at 70 Pa, about 7×10^{21} and 6×10^{21} m^{-3}, respectively. At double the P_{EFF}, these n_O values are approximately double compared to the n_O values at $P_N = 100$ W. The n_O at $P_N = 300$ W is slightly less than thrice the n_O at $P_N = 100$ W.

The n_O also appears to increase with decreasing working pressure, at least at $P_N = 200$ W. As explained above, the larger mean free path at low pressures increases the electron temperature, and thereby, the electron density via higher ionization probability (Cvelbar et al. 2007, 224-227). As more electrons collide with O_2 molecules, production of O via dissociation increases, which results in an increased n_O.

At P_N = 100 W, n_O is recorded at a pressure of 30, 50, and 70 Pa. At P_N = 200 W, it is only recorded at 50 and 70 Pa, and at P_N = 300 W, only at 70 Pa. This is because the plasma assumes the H-mode at the remaining conditions in the absence of a garlic sample.

6.3 Probe signal increase during clove treatment

When a garlic sample is introduced into the plasma, the recorded catalytic probe signal increases compared to plasma in an empty reactor. This is shown in Figure 6.3, a comparison of n_O recorded during E-mode plasma alone and during treatment of a peeled clove. The initial n_O value during sample treatment is equal or larger than for plasma alone, followed by a gradual increase until a plateau is reached, whereas the n_O for plasma alone remains relatively constant throughout the experiment.

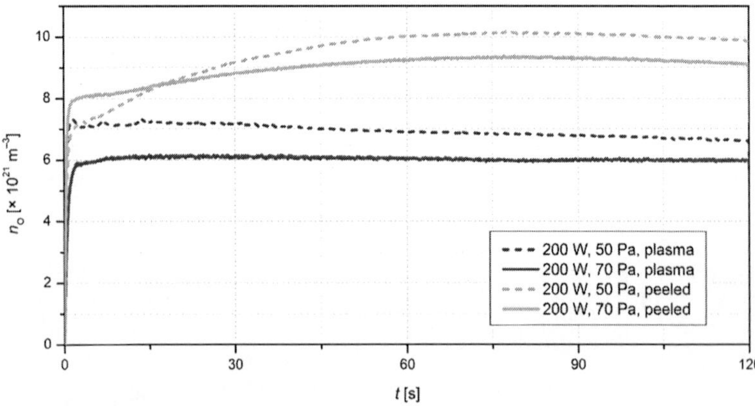

Figure 6.3: n_O, pilot-scale E-mode, plasma or peeled clove. P_N = 200 W, p = 50 – 70 Pa, t = 120 s. The discharge is turned on at t = 0 s and turned off at t = 120 s.

The increase we observe is contrary to what is expected for polymers or other materials that are functionalized by oxygen plasma. In a plasma system, the oxygen atom density depends on the equilibrium between the production and loss of the atoms. If a material treated in a plasma is functionalized, the n_O should decrease, as the oxygen atoms are "spent" introducing new functional groups to the surface.

To understand this observation, we must consider how the n_O is calculated. The signal detected by the catalytic probe depends on the amount of recombinations of reactive plasma species that occur at the probe tip surface (Šourková et al. 2018). To calculate n_O, we assume that the signal increase is due to oxygen atom recombinations ($O + O \rightarrow O_2$) only, which is true for plasma alone if the effect of ions is excluded. However, the probe is not specific to O recombinations, as other exothermic reactions also affect the signal. This includes recombinations of OH ($OH + H \rightarrow H_2O$) and CO ($CO + O \rightarrow CO_2$) radicals.

From this reasoning, it follows that the signal increase recorded in the presence of garlic does not represent the recombination of O atoms alone. As explained above, etching and water loss from the clove introduce new species into the gaseous phase during plasma treatment of cloves. The OH and CO radicals released in this way then recombine at the probe tip surface and increase the recorded signal. Thus, Figure 6.3 confirms the occurrence of etching and water desorption from the clove.

Further, in Figure 6.4, we compare the signal increase between plasma alone, treatment of one or three peeled cloves, and treatment of one or three unpeeled cloves. We see that the recorded signal is higher in all cases of garlic sample treatment. Further, it is higher during the treatment of three cloves compared to one clove, as well as during the treatment of peeled cloves compared to unpeeled cloves. This is explained by the amount of CO and OH radicals released from the cloves. A higher number of cloves has a larger surface exposed to the plasma than a single clove, and thus releases more CO and OH during treatment. Similarly, the peeled clove is a practically limitless supply of water to serve as a source of OH radicals, while the water content in the protective leaf is rather low, and thus likely exhausted soon after the beginning of treatment.

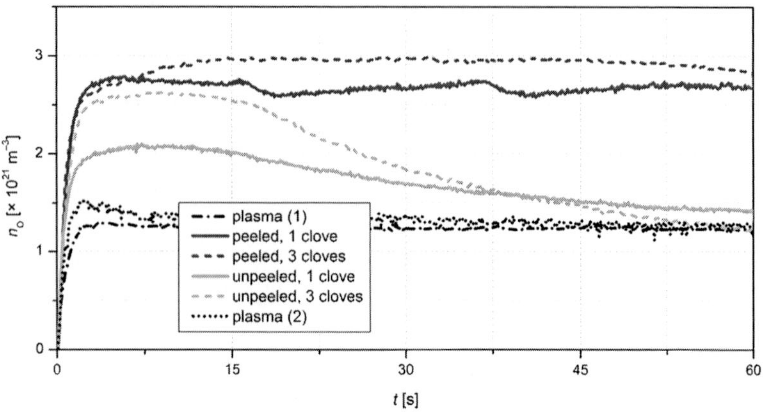

Figure 6.4: n_O, pilot-scale E-mode, plasma or peeled or unpeeled clove. $P_N = 100$ W, $p = 30$ Pa, $t = 60$ s. Cloves are treated either individually or as three cloves simultaneously. The discharge is turned on at $t = 0$ s and turned off at $t = 60$ s.

6.4 Treatment of peeled cloves

The probe signal recorded during treatment of a peeled clove increases with increasing P_N, as well as with decreasing working pressure, as shown in Figure 6.5 (A). The increase is similar to what we see in plasma alone. However, in addition to the equilibrium of O atom formation and loss, the introduction of OH and CO into the system must be accounted for in the explanation of signal behavior. We see that the signal increases during the course of the treatment become steeper with increasing P_N. This is explained by the intensity of etching and water loss, which is higher at increased P_N. The intensity of the OH line in OES spectra indicates the importance of the OH radical in shaping the signal recorded by the catalytic probe. The OH signal in the OES spectrum is about 1.5 × higher for $P_N = 300$ W compared to $P_N = 200$ W, and so is the signal recorded at the catalytic probe.

From Figure 6.5 (B), we see that the plateau reached during short treatment (< 120 s) gradually begins to drop once the treatment time crosses ~ 100 s, and continues to do so during the 300 s of treatment. This is likely due to the exhaustion of the water adsorbed on the surface and found in the upper surface layers. Water from deeper layers of the clove is more difficult to bring to the surface during treatment, and takes a longer time to do so. Thus, with the density of OH radicals slowly dropping, the signal at the catalytic probe begins to gradually decrease.

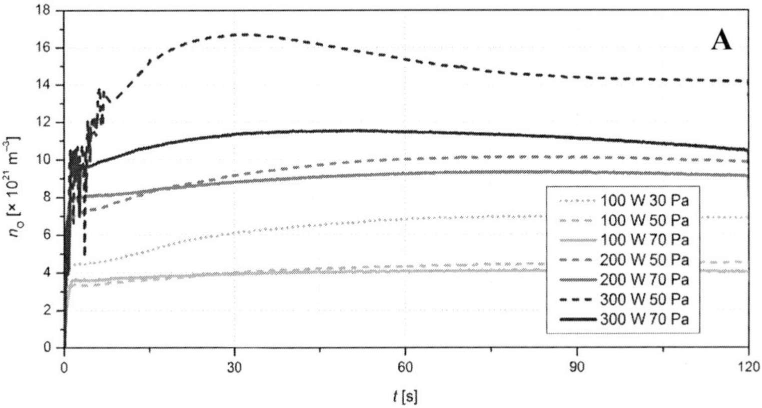

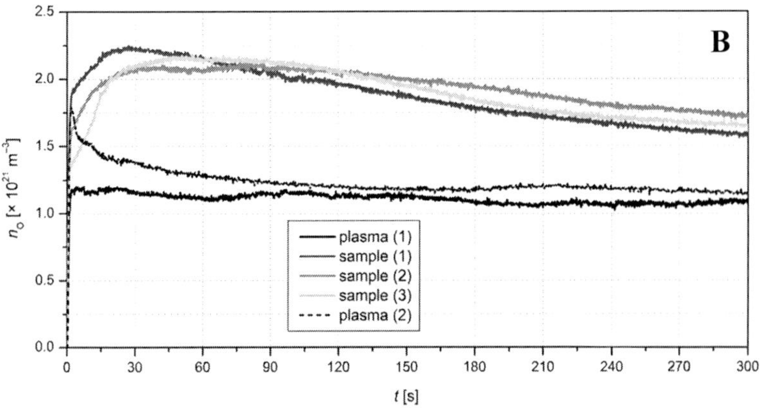

Figure 6.5: n_O during treatment of peeled garlic, E-mode glow. (A) P_N = 100 – 300 W, p = 30 – 70 Pa, t = 120 s. (B) P_N = 100 W, p = 30 Pa, t = 300 s.

In Figure 6.6, we see that the n_O during H-mode afterglow treatment of a single peeled clove at the given conditions is about 2.5 – 3 × 10^{21} m^{-3}. Compared to the measurements in Figure 6.5 (A), this value is quite modest. In addition to a smaller n_O due to the distance from the coil, this is also a consequence of less intense surface etching, and thereby less intense water loss, in the afterglow position.

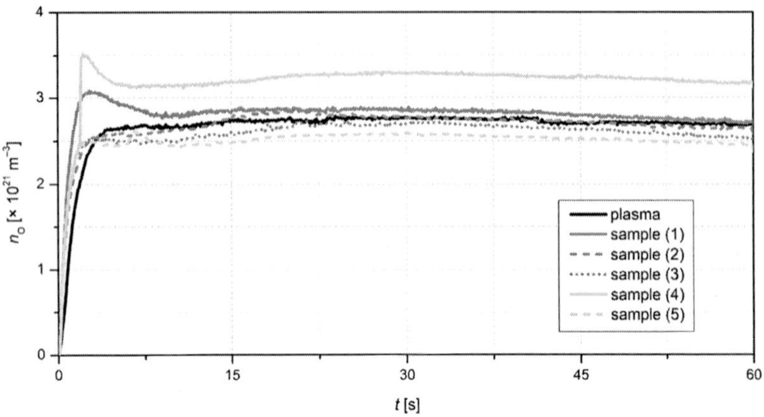

Figure 6.6: n_O during treatment of peeled garlic, H-mode afterglow. P_N = 300 W, p = 30 Pa, t = 60 s. The discharge is turned on at t = 0 s and turned off at t = 60 s.

6.5 Treatment of unpeeled cloves

As above, Figure 6.7 (A) shows that the recorded probe signal during treatment of a peeled clove increases with increasing P_N and decreasing working pressure. The irregular signal obtained at 300 W, 50 Pa is the consequence of burning of the exposed edges of the protective leaf.

It is interesting that during prolonged treatment of unpeeled cloves in E-mode plasma, shown in Figure 6.7 (B), the signal on the catalytic probe rises to a peak at about 15 s, falls to a low plateau at about 100 s, and begins to rise again at about 200 s. The initial peak may present either loss of water from the protective leaf (which is more plausible), or intense initial etching (which is less likely due to the low P_{EFF} and the absence of etching products in OES spectra at these plasma parameters). As the protective leaf is dryer than the clove itself, it is possible that most of its water is lost in the first minute, after which, there is a period of functionalization, followed by etching after about 200 s.

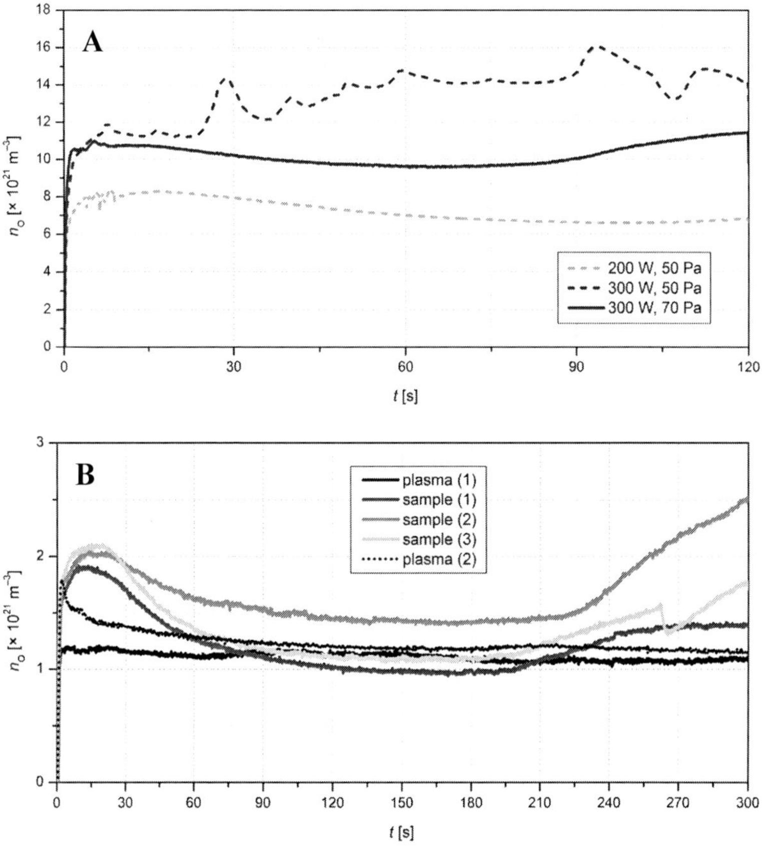

Figure 6.7: n_O during treatment of unpeeled garlic, E-mode glow. (A) $P_N = 200 - 300$ W, $p = 50 - 70$ Pa, $t = 120$ s. (B) $P_N = 100$ W, $p = 30$ Pa, $t = 300$ s.

As above, we see in Figure 6.8 that the n_O during H-mode afterglow treatment of a single unpeeled clove at the given conditions is about $2.5 - 3 \times 10^{21}$ m^{-3}. The value is again a result of the distance from the coil, less intense surface etching, and less intense water loss.

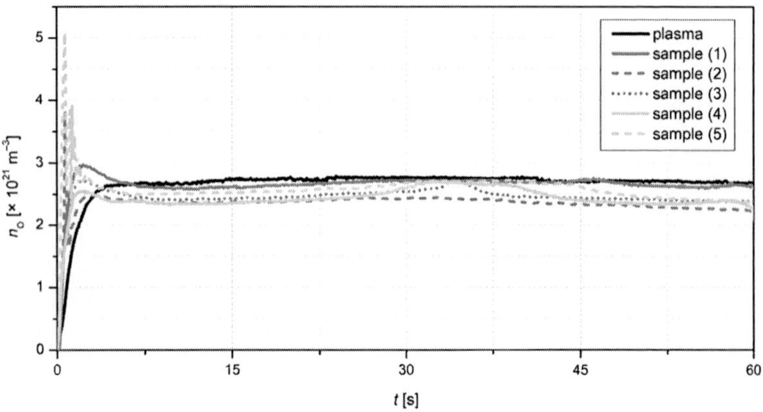

Figure 6.8: n_O during treatment of unpeeled garlic, H-mode afterglow. P_N = 300 W, p = 30 Pa, t = 60 s. The discharge is turned on at t = 0 s and turned off at t = 60 s.

7. TEMPERATURE

7.1 Thermocouple measurements

During exposure of a sample to a plasma, its temperature increases as a result of reactive species flux to the sample surface. For the treatment of delicate organic materials such as garlic cloves, it is important to select discharge parameters where the temperature increase does not negatively affect the sample. To ensure this, sample temperature during plasma treatment should ideally be recorded as well.

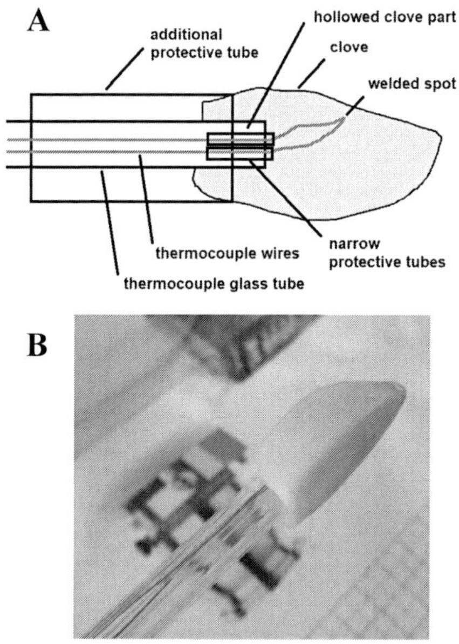

Figure 7.1: Thermocouple measurements. (A) Schematic representation of mounting a thermocouple inside a garlic clove for temperature measurement; (B) a clove with the inserted thermocouple before treatment.

One way to record the temperature of garlic cloves during plasma treatment is to mount a simple thermocouple (TC) inside the clove. This is shown schematically in Figure 7.1. In this particular instance, a couple of TC wires from different conductors are spot welded on one side. Cloves are cut at the bottom and partly hollowed out to allow access, and the welded spot is inserted into the storage leaf a few mm under the surface of the clove, assuring no electrical contact between the wires except at the welded spot. The wires stretching from the clove are insulated with narrow glass tubes, and an additional glass tube is covering the bottom part of the clove. On the other side, the wires are connected to a voltage meter to measure the temperature-dependent voltage produced as a result of the thermoelectric effect; the data is collected and interpreted as temperature measurements.

7.2 Initial heating

To evaluate the effect of heating during plasma treatment on garlic cloves, let us examine TC measurements at selected discharge parameters. Figure 7.2 follows exposure of a peeled clove with an inserted TC, or a TC alone, to pilot-scale E-mode glow plasma with P_N = 200 W. As expected, the temperature starts increasing upon turning on the discharge, which is explained by exothermic surface effects rather than immediate increase of the gas temperature. Namely, the neutral gas kinetic temperature in oxygen RF plasma sustained with an electrodeless discharge created in a glass tube at low pressure and power remains practically at room temperature due to the lack of gas-phase collisions that would cause transfer of available power to the kinetic energy of gaseous molecules (Doliška et al. 2012, 56-61). We further see that the temperature progression at a working pressure of 50 Pa, when the discharge is ignited, and at atmospheric pressure, where there is no discharge, is nearly identical. A TC without a garlic sample reaches twice the temperature of the garlic measurement within the first minute, but settles at a plateau thereafter.

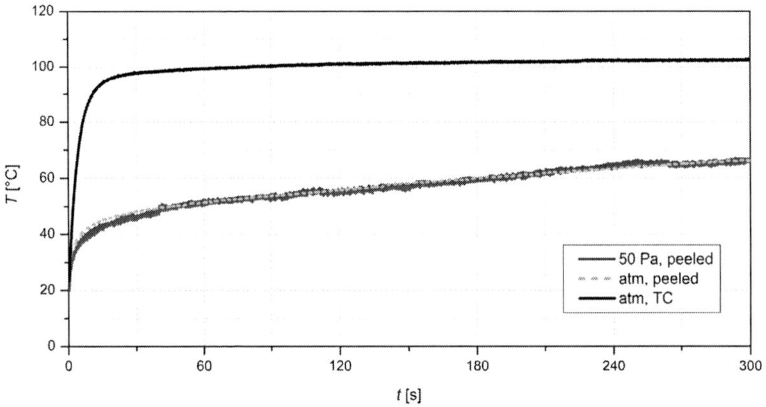

Figure 7.2: Temperature measured by a TC in the pilot-scale plasma P_N = 200 W, p = 50 Pa or atm, t = 300 s, sample = peeled garlic or TC alone.

The initial stage of heating (for the first 10 seconds of plasma treatment) at different discharge conditions is shown in Figure 7.3. The initial temperature is about 25 – 30 °C. Just after turning on the discharge, the temperature for all four cloves drops for 10 °C or more. There is no reason for immediate cooling of a clove in a fraction of a second, so this effect is explained by radio frequency interferences. Although both the thermocouple wires and the voltage meter are kept at floating potential, there is some interference, so the sudden drop of the temperature just after turning on the discharge is an artifact. Following that, the measured temperature during plasma treatment of a peeled clove increases rather intensely for the first few s. It rises from about 20 °C to about 40 °C within the first 10 s of treatment. After this initial rise, the temperature assumes a rather slow increase.

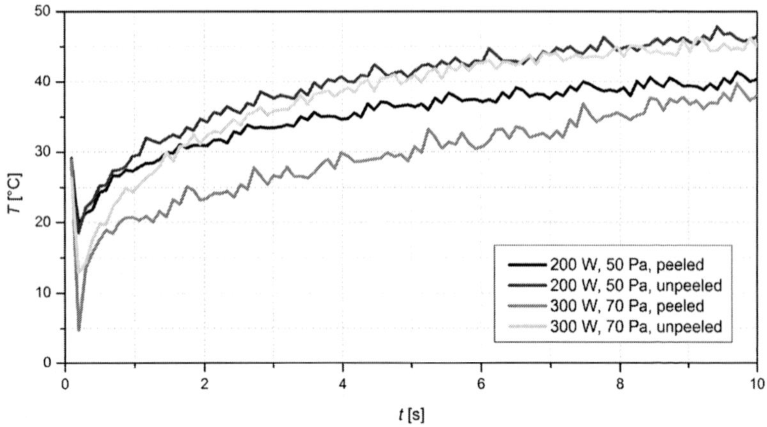

Figure 7.3: T time course, peeled and unpeeled cloves, pilot-scale plasma. $P_N = 200 - 300$ W, $p = 50 - 70$ Pa, sample = peeled or unpeeled clove; the first 10 s of treatment.

Neglecting the interference artifact, it is possible to explain the heating of the cloves after turning on the discharge. If the heating power were constant and the thermal conductivity of the cloves infinite, the initial increase of the clove temperature would have been linear with time for the first second after turning on the discharge. Namely, after turning on the discharge, any cooling effects are easily neglected since the temperature difference between the clove and surrounding gas is marginal. The linear approximation of the temperature curves after turning on the discharge enables determination of the heating power. At the conditions presented in Figure 7.3, the time derivative of a clove is about 15 K/s. One can calculate the power taking into account a standard relation, i.e.

$$P = m\,c_\mathrm{p}\,\frac{\Delta T}{\Delta t}. \tag{7.1}$$

Here, P is the heating power, m the mass of the clove, c_p the specific thermal capacity of garlic at room temperature, and $\Delta T/\Delta t$ the slope of the linear approximation of the temperature curve after turning on the discharge. The specific thermal capacity of garlic is unknown, but can be replaced with a value typical for fresh fruit, i.e., 3.5×10^3 J kg^{-1} K^{-1} (Engineering ToolBox 2003). This assumption is reasonable since the major component (~ 65 %) of a garlic clove is water, same as for fresh fruit. The rest is biological matter (complex hydrocarbons of lower thermal capacity), so the specific heat capacity is somewhat lower than for pure water (4.2×10^3 J kg^{-1} K^{-1}). Still, one can take into account the numerical values using available data and calculate the heat dissipated on a clove surface in unit time after turning on the discharge using Eq. (7.1). The mass of a clove was about 1 g and the time derivative about 15 K/s, so the initial heating power was

$$P = 10^{-3}\ \mathrm{kg} \times 3.5 \times 10^{3}\ \mathrm{J\ kg^{-1}\ K^{-1}} \times 15\ \mathrm{K\ s^{-1}} = 50\ W. \quad (7.2)$$

From Table 4.1, we know that the P_{EFF} at these discharge conditions is 50 W. If the above calculation is correct, this would mean that all available discharge power is used for heating of the clove. As shown below, other processes, namely functionalization and etching, also take place, so this is obviously untrue. In Figure 7.2, we additionally see that at the same experimental conditions, but in atmospheric air instead at 50 Pa of oxygen, the temperature time course is practically identical than in the above experiment. At atmospheric pressure, no discharge appears when the generator is turned on. This means that the temperature increase in this case is due to induction heating, as the sample is exposed to the EM field formed by the coil. Therefore, we can state that the temperature increase of garlic during plasma treatment is not as high as suggested by the TC measurements.

As explained above, the power used for clove heating is dissipated on the surface of the clove, so the relevant parameter is the power per unit surface. Cloves are not geometrically simple, but a useful approximation of the surface area is to take into account spherical geometry. Such an approximation underestimates the real surface of a clove, but on the other hand, a significant part of clove surface is shaded from plasma since the clove is placed onto a microscope slide during treatment. The surface of a sphere is calculated as

$$A = 4\pi\, (\frac{3m}{4\pi\rho})^{2/3}. \quad (7.3)$$

Here, ρ is the density of garlic. In a reasonable approximation, the density is close to the water density, i.e., $\rho = 10^3$ kg m^{-3}. The surface of a garlic clove with a mass of 1 g facing plasma is therefore close to

$$A = 13\, (\frac{3 \times 10^{-3}\ \mathrm{kg}}{13 \times 10^{3}\ \mathrm{kg\ m^{-3}}})^{2/3} = 13\, (0.23 \times 10^{-6}\ \mathrm{m^{-3}})^{2/3} \\ = 5 \times 10^{-4}\ \mathrm{m^{-2}} = 5\ \mathrm{cm^2}. \quad (7.4)$$

The power density (i.e., power per unit surface area, as determined from the measured temperature from Figure 7.3) is therefore around 10^5 Wm^{-2} = 0.1 W mm^{-2}:

$$\frac{P}{A} = \frac{50\ \mathrm{W}}{5 \times 10^{-4}\ \mathrm{m}^{-2}} = 1 \times 10^5\ \mathrm{W\ m^{-2}}. \qquad (7.5)$$

Gaseous plasma was characterized by electrical and catalytic probes. The density of oxygen ions at the position of the sample was previously estimated using a double electrical probe and the value was about $n_\mathrm{i} = 8 \times 10^{15}$ m^{-3}. The density of neutral oxygen atoms at the same position as determined with a fiber-optics catalytic probe (Figure 6.2) was $n_\mathrm{O} = 7 \times 10^{21}$ m^{-3} at these discharge conditions.

The effects of weakly ionized highly reactive oxygen plasma are highly surface specific. There are two dominant heating mechanisms of solid materials facing this kind of plasma: (1) heating by positively charged molecular ions due to both surface neutralization and bombardment, and (2) heterogeneous surface recombination of oxygen atoms to parent molecules. Other effects like absorption of light quanta, surface relaxation of metastables, accommodation of any vibrational-excited molecules, and exothermic chemical reactions (low temperature “burning”) are regarded marginal for such plasma, as explained in detail elsewhere (Gorjanc and Mozetič 2014).

Let us first consider the heating of cloves by positively charged ions. The ions move randomly in bulk plasma and their temperature is close to the neutral gas kinetic temperature, i.e., 300 K. Once approaching a surface, they first enter the pre-sheath, where they assume directed velocity which is calculated as the "Bohm velocity", i.e.

$$v_\mathrm{Bohm} = \sqrt{\frac{k_\mathrm{B} T_\mathrm{e}}{M}}. \qquad (7.6)$$

Here, k_B is the Boltzmann constant, T_e the electron temperature in plasma, and M the ion mass. Taking into account the typical electron temperature in weakly ionized oxygen plasma, 5×10^4 K, and the ion mass, 32 Da, the Bohm velocity is

$$v_\mathrm{Bohm} = 3 \times 10^3\ \mathrm{m\ s^{-1}}. \qquad (7.7)$$

The positively charged ions are accelerated in the sheath next to any floating object immersed into gaseous plasma. In the collision-less approximation, which is justified for such a plasma at the pressure of 50 Pa, the power dissipated on the surface of a garlic clove per unit surface area is

$$\frac{P_{\text{ion}}}{A} = j_{\text{ion}} W_{\text{ion}}, \tag{7.8}$$

where j_{ion} is the flux of ions towards the surface (product of ion density and Bohm velocity), and W_{ion} the available energy, i.e., the sum of potential and kinetic energies. Both energies are close to 10 eV, so one can use the value $W = 20$ eV $= 3 \times 10^{-18}$ J. Inserting numerical values, i.e. $n_{\text{i}} = 8 \times 10^{15}$ m^{-3}, $j_{\text{ion}} = 2.4 \times 10^{19}$ m^{-2} s^{-1} and $W_{\text{ion}} = 3 \times 10^{-18}$ J, one calculates the power dissipated per unit area on the surface of a garlic clove due to interaction of positively charged molecular ions as

$$\frac{P_{\text{ion}}}{A} = 2.4 \times 10^{19}\ \text{m}^{-2}\ \text{s}^{-1} \times 3 \times 10^{18}\ \text{J} = 72\ \text{W m}^{-2}. \tag{7.9}$$

The value in Eq. (7.9) is three orders of magnitude smaller than the measured power density presented in Eq. (7.5). Although several approximations were used to estimate the heating of garlic cloves by positively charged ions, it is clear that the ions play a marginal role.

We have established that the density of neutral oxygen atoms in weakly ionized oxygen plasma is orders of magnitude larger than the ion density, around $n_{\text{O}} = 7 \times 10^{21}$ m^{-3} in the pilot-scale reactor experiment at the given parameters (Figure 6.2). The atoms are neutral, so they are not accelerated in the sheath next to the sample surface, but move thermally, so the flux of atoms onto the surface of any object facing plasma is

$$j_{\text{O}} = \frac{1}{4} n_{\text{O}} < v_{\text{a}} >. \tag{7.10}$$

Here, $<v_{\text{O}}>$ is the average velocity of O-atom thermal motion, which is calculated from Boltzmann law as

$$< v_{\text{O}} > = \sqrt{\frac{8 k_{\text{B}} T}{\pi \frac{M}{2}}}. \tag{7.11}$$

Taking into account numerical values and assuming a room temperature of $T = 300$ K, one gets the average random velocity of O-atoms

$$< v_{\text{O}} > = 630\ \text{m s}^{-1}. \tag{7.12}$$

The heating due to interaction of O-atoms with the garlic surface is calculated using Eq. (7.8), except that W from Eq. (7.8) should be replaced with "average energy an atom brings to the surface". If all atoms recombine to molecules, and the molecules leaving the surface are in thermal equilibrium with the solid material, the heating by heterogeneous surface recombination of O-atoms to parent molecules will be

$$\frac{P_{\mathrm{O}}}{A} = \frac{1}{2}\, j_{\mathrm{O}} W_{\mathrm{d}}. \tag{7.13}$$

Here, W_{d} is the dissociation energy of the oxygen molecule: $W_{\mathrm{d}} = 5.2\ \mathrm{eV} = 8 \times 10^{-19}\ \mathrm{J}$. The factor 1/2 comes from the fact that an oxygen molecule contains 2 atoms. The heating by atoms in the approximation of 100 % surface recombination would be

$$\begin{aligned}\frac{P_{\mathrm{O}}}{A} &= \frac{1}{8}\, 7 \times 10^{21}\ \mathrm{m^{-3}} \times 6.3 \times 10^{2}\ \mathrm{m\,s^{-1}} \times 8 \times 10^{-19}\ \mathrm{J} \\ &= 4 \times 10^{5}\ \mathrm{W\,m^{-2}}.\end{aligned} \tag{7.14}$$

The calculated value is larger than the measured heating power as shown in Eq. (7.5) by a factor of 4. The discrepancy is explained by the fact that the surface recombination probability is not 100%. As mentioned earlier, the recombination coefficient depends on the type of material and its morphology (Mozetič et al. 2015, 015026). The value as deduced from Eq. (7.5) and (7.14) is surprisingly large and worth discussing. Namely, dividing the calculated value in (7.14) with the measured one in (7.5) gives a recombination probability of

$$\gamma = 0.25. \tag{7.15}$$

This value is typical for smooth catalytic materials and definitely well above the order of 10^{-3} typical for many polymers (Zaplotnik et al. 2018, 1800021). Such a large recombination coefficient could be explained by several effects, the most obvious of which are (1) surface roughness, (2) the presence of catalytic materials on the surface of an object facing plasma. The surface roughness plays an important role in surface recombination of oxygen atoms. In fact, the largest recombination coefficients exceeding the value $\gamma = 0.5$ have been measured for a rather inert material of extremely rich morphology on the sub-micron scale (Mozetič et al. 2015, 207-213). The effect was explained by trapping O-atoms into pores of dimension well below the mean free path of O-atoms, which is roughly 200 μm at a pressure of 50 Pa. In the SEM images of untreated peeled clove surface shown in Figure 9.5, we see that the surface is rather smooth, and despite its roughening with increased treatment time, certainly not rough enough to cause such trapping of O atoms. From this consideration, one can conclude that the surface roughness is not sufficient to explain the recombination coefficient presented in Eq. (7.15). The surface composition of an untreated peeled garlic clove as determined by XPS is shown in Table 8.1. The surface is composed predominantly of hydrocarbons with a few at.% of oxygen and a few other elements in marginal amounts. There is obviously no catalytic material on the surface of untreated garlic. This confirms our earlier conclusion that the calculated heating power is overestimated due to the effect of induction heating on the recorded temperature values. As such, the TC measurements cannot be used to determine the absolute temperature of the treated clove, but a relative comparison is still possible.

7.3 Later heating

After several seconds of plasma treatment, the increase of the garlic temperature is not as extensive as initially. Figure 7.4 reveals that the temperature still increases with increasing time for the first few minutes, but the increase is slower. If the curve from about 30 to 300 s is interpolated with a line, the temperature increase in this region will be a bit below 0.1 K/s. Such a slow increase could be explained by several effects, including the slow modification of the surface morphology, weak variation of the recombination coefficient, or increased temperature of gas surrounding the clove. The latter explanation is preferred. Such a slow increase is typical for thermal effects, since the entire discharge tube slowly warms up, allowing gas molecules to accommodate on warm surfaces and thus causing an increase of the gas temperature.

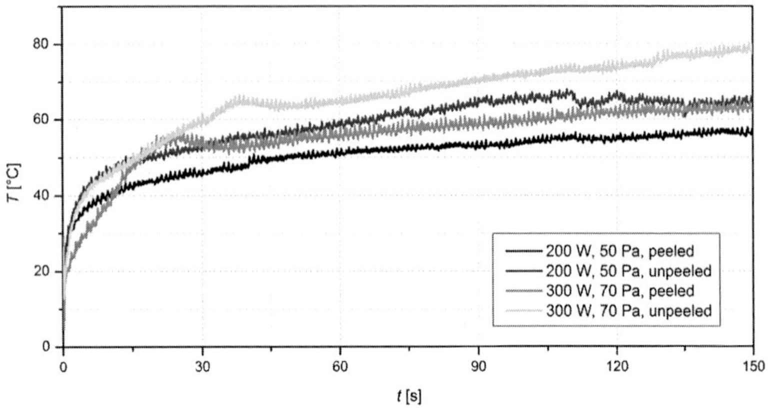

Figure 7.4: T time course, peeled and unpeeled cloves, pilot-scale plasma. $P_N = 200 - 300$ W, $p = 50 - 70$ Pa, sample = peeled or unpeeled clove; treatment up to $t = 150$ s.

7.4 Heating at different discharge conditions

Figure 7.4 shows the time course of temperature during peeled and unpeeled clove treatment at two different discharge conditions: 200 W at 50 Pa, as well as 300 W at 70 Pa. As explained above, these temperature values are not absolute due to induction heating effects, but can be used for relative comparison. In Figure 6.2, we see that the oxygen atom density is around 7×10^{21} m^{-3} at 200 W and 50 Pa, while it is around 8.5×10^{21} m^{-3} at 300 W and 70 Pa. Therefore, it is unsurprising that the clove treated at the higher P_N heats slightly more rapidly than the one treated at the lower P_N.

It is further interesting to note that unpeeled cloves heat somewhat more rapidly than peeled cloves. This is due to the nature and function of the protective leaf, which prevents water and heat from leaving the clove. For this reason, heat damage affects unpeeled cloves earlier, i.e., at shorter treatment times than peeled cloves, as will be shown below in the sprout and root growth section.

8. SURFACE CHEMISTRY

8.1 Plasma effects on surface chemistry

During exposure of garlic cloves to plasma, the reactive species of the plasma directly affect one particular segment of the clove: its surface. In general, non-thermal plasma is well known for its use in surface modification, including tuning the properties of organic materials such as polymers (Gorjanc and Mozetič 2014). It is a very surface-specific technique; depending on treatment parameters and duration, it typically affects a smooth surface up to some tens of nm in depth (Molina et al. 2018). The result is modification of physico-chemical properties of the plasma-treated surface.

One important aspect of these modifications is functionalization – the introduction of new functional groups onto the surface. It is established that treatment with oxygen-containing plasma will initiate oxidation processes on the treated material (Gorjanc and Mozetič 2014; Wheale et al. 1998, 6699-6704), introducing new polar functional groups and oxidizing existing ones (Gorjanc and Mozetič 2014; Zaplotnik et al. 2016, 144; Mozetič 2010, 165-171) through grafting of oxygen atoms onto the surface of the material (Belmonte et al. 2005, 26-30). These oxygen-containing functional groups can include hydroxyl, ether, carbonyl, ester, and carboxyl groups, as well as unstable groups oversaturated with oxygen (Gorjanc and Mozetič 2014).

8.2 Methods of analysis

Surface chemistry of organic materials can be analyzed by a variety of techniques. A suitable one for the analysis of very thin layers of the surface is X-ray photoelectron spectroscopy (XPS). It uses an X-ray beam to irradiate the material surface, and detects the number and kinetic energy of electrons escaping from that surface. From this data, number of electrons detected may be plotted versus their binding energy, which gives information about elemental composition, expressed in atomic percent (at.%). The detection is limited to the outermost surface layer (typically about 5 nm). In this way, changes to the surface elemental composition due to introduction of new chemical groups, or removal of layers of material from the surface, can be detected. Further analysis includes fitting of sub-peaks to the peaks of high-energy resolution spectra, which provides additional information regarding the chemical bond composition of the sample surface. XPS analysis requires placing the sample in high or ultra-high vacuum conditions. In the case of garlic, this means that unpeeled and peeled cloves need to be treated differently. To analyze unpeeled cloves, the protective leaf may be removed after treatment, and is dry enough to be analyzed directly. To analyze peeled cloves, pieces excised from the clove surface need to be freeze dried before analysis.

If an even more surface sensitive technique is needed, Time of Flight Secondary Ion Mass spectrometry (ToF-SIMS) can be employed, as it provides information about the outermost surface layer (about 1 nm) with high lateral resolution (50 – 100 nm). It enables analysis of surface composition of solid surfaces and thin films, specifically identification of molecular structure. In ToF-SIMS, the surface is bombarded with a focused high energy (1 – 30 keV) ion beam. This causes emission of species from the solid surface. Together with other particles, the so-called secondary ions are emitted from the surface, after which, mass spectrometry imaging (MSI) is employed to detect and analyze them. As with XPS, SIMS also uses ultra-high vacuum conditions to operate. Therefore, in the case of biological samples, sample preparation is a crucial step required to maintain the natural state of biomolecules prior to analysis. Freeze-drying methods are most commonly used, while frozen hydrate analysis is the preferred sample preparation method (Yoon and Lee 2018, 24). The advantage of SIMS is that it can provide valuable information on trace elements present at very low concentrations, even in the ppb range.

Another useful technique is infrared (IR) spectroscopy. In the case of attenuated total reflectance Fourier-transform infrared (ATR-FTIR) spectroscopy, an IR beam passes through the sample and becomes attenuated in regions of the IR spectrum where the sample absorbs energy. In this way, additional information about surface chemistry is obtained – in particular, information about the chemical bonds present at the surface. The result is an IR absorption spectrum which plots transmittance (%) against wavenumber (cm^{-1}). The spectra are roughly divided in two regions: the functional group region (4000 – 1500 cm^{-1}), where peaks correspond to specific chemical bonds, and the fingerprint region (1500 – 380 cm^{-1}), where peaks correspond to complex molecular deformations. In contrast to XPS, the analysis depth of IR spectroscopy is much larger, typically between 0.5 and 2 μm.

8.3 Untreated clove surface chemistry

As stated, XPS is a very surface-sensitive technique, analyzing the chemical composition of the surface about 5 nm in depth. Given that plant cuticle thickness is typically between 0.1 and 10 μm (Riederer and Müller 2006), XPS only analyzes a thin outermost layer of the cuticle – specifically, the epicuticular waxes coating its outer surface. If we compare the unpeeled and peeled garlic surfaces, it becomes clear that the epicuticular wax compositions of the protective leaf and the clove surface differ somewhat, which is commonly the case between different plant organs in the same species (Riederer and Müller 2006).

Table 8.1 summarizes the surface elemental compositions of the untreated garlic clove surfaces from various experiments, both for the protective leaf and the clove surface. The results show that these surfaces consist mainly of C and O, with minor amounts of other elements that differ by experiment. The oxygen concentration is higher at the protective leaf surface than at the peeled clove surface. However, the protective leaf surface composition and O/C ratio also appear to vary in a wider range than those of the peeled clove surface. The mean O/C ratio is lower on the clove surface compared to the protective leaf surface.

Table 8.1: Surface analysis of untreated clove surfaces. Elemental composition [at.%] and O/C ratios. Experiments: (PL) protective leaf, (C) clove. (PS) pilot-scale, (IS) industrial-scale. (G) glow, (AG) afterglow.

experiment	**C**	**O**	**Ca**	**Si**	**N**	**Cl**	**Na**	**O/C ratio**
PL, PS, G	90.2	8.2	0.1	/	1.4	0.1	/	0.091
PL, PS, AG	96.6	3.1	/	/	/	/	/	0.032
PL, PS pulsed	93.9	4.7	/	0.8	0.6	/	/	0.050
C, PS G	97.5	2.5	/	/	/	/	/	0.026
C, IS G	97.8	2.2	/	/	1.3	0.1	0.6	0.022

The higher oxygen concentration at the protective leaf surface compared to the clove surface indicates a higher degree of innate oxygenation of the protective leaf cuticular waxes. As shown later in ATR-FTIR spectra (Figure 8.14), the oxygen is mainly present in the form of –OH and C=O groups, which are typically present as terminal functional groups of cuticular wax compounds (Riederer and Müller 2006).

Further analysis shows that the wettability of the native protective leaf and clove surfaces is comparable (Figure 10.1), while the water uptake of untreated unpeeled cloves is much higher than that of untreated peeled cloves (Figure 10.2). This indicates a higher hydrophobicity of the native clove surface compared to that of the native protective leaf surface, and the lower O/C ratio at the clove surface is therefore expected. This indicates that the peeled clove cuticular waxes contain a higher fraction of non-functionalized hydrocarbons, which are more hydrophobic than their equivalents with polar, oxygen-containing functional groups.

The greater variability of C and O content in the protective leaf may be attributed to biological variability, which is presumably higher in the protective leaf than the clove itself. Often, variation in the protective leaf of different cloves, from the same garlic batch or even the same bulb, is already apparent in the protective leaf color, thickness, or texture. On the other hand, all healthy peeled cloves appear uniformly white and smooth.

To the best of our knowledge, the only XPS analyses of garlic in available literature have been performed on garlic peel that was processed and ground before analysis, with emphasis on the detection of externally added adsorbed materials (Liu et al. 2014, 2054-2063; Zhao et al. 2018, 15264-15272). On the other hand, instances of XPS analyses of other native plant cuticles are found. In tomato fruit cuticles, 96.4 % C and 3.6 % O were detected on the cuticle surface, corresponding to an O/C ratio of 0.037 (Heredia-Guerrero et al. 2012, 803-808). On the adaxial leaf surfaces of two clones of willow, the concentration of C, O and N was either 86.4, 12.0 and 1.6 %, or 89.7, 8.8 and 1.5 % (Hietala et al. 1997, 205-215). Cherry laurel leaf surfaces showed an average of 94.0 % C, 4.7 % O, 0.9 % N, and 0.4 % Si (Perkins et al. 2005, 123-134). This confirms that plant cuticle waxes vary in their surface chemical composition, but nonetheless consist mainly of hydrocarbons with some oxygen-containing functional groups (Riederer and Müller 2006).

8.4 Carbon and oxygen content

8.4.1 Protective leaf

8.4.1.1 E-mode glow treatment

After oxygen plasma treatment of organic materials, in particular aliphatic hydrocarbons, oxygen concentration at the surface typically increases with increased dose of plasma reactive species (Gerenser 1987, 303-318; Øiseth et al. 2002, 92-103). This is true for all results of XPS measurements performed on the garlic protective leaf, where increased treatment time and/or generator power has resulted in an increased oxygen concentration and an increased O/C ratio. Surface C and O concentrations of the garlic protective leaf are shown in Figure 8.1, and the O/C ratio is shown in Figure 8.2. During glow plasma treatment, carbon content at the surface decreases, and oxygen content increases, with increased treatment time. The O/C ratio increases nearly linearly with glow plasma treatment time. Compared to glow treatment, the increase with the afterglow treatment is less steep.

We see a notable decrease in C and increase in O content after as little as 5 s of treatment. The fastest increase in surface O content is achieved within the first 30 s of treatment, and with increasing treatment time, the O content rise becomes more gradual. Saturation of organic material surface with oxygen-containing functional groups is typical, and is complemented by chemical etching, which is further explained below. Within the treatment times used here, saturation of the clove surface has not been completely achieved, but is gradually approached. This behavior could also be the consequence of simultaneous functionalization and etching, during which, organic material is removed from the surface, deeper layers of varying chemical composition are uncovered, and thus analyzed with XPS at the given treatment time.

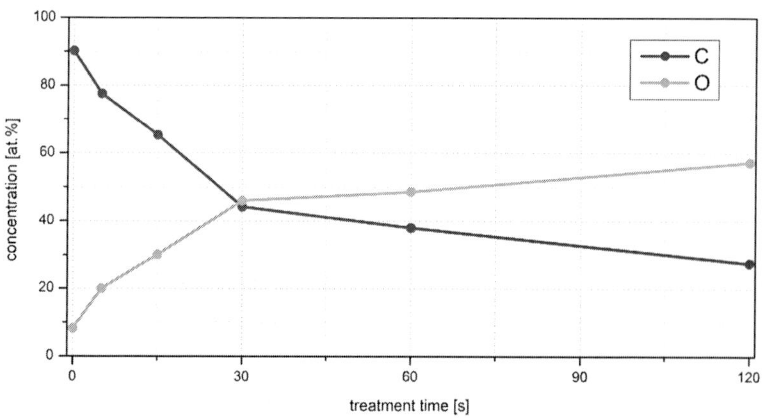

Figure 8.1: C and O concentration at the protective leaf surface. Pilot-scale E-mode glow, $P_N = 200$ W, $p = 50$ Pa, $t = 0 - 120$ s.

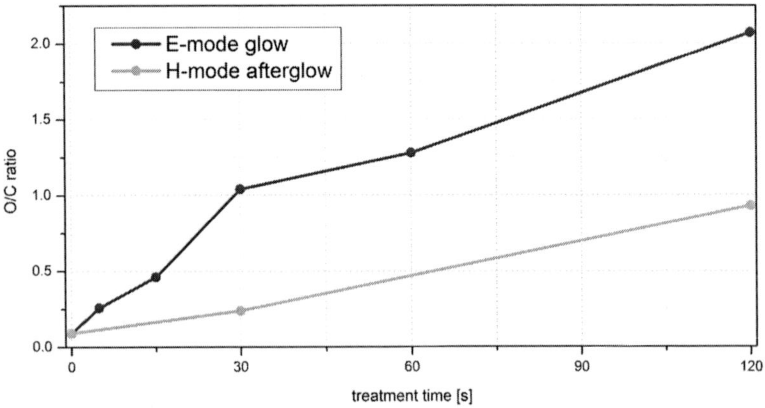

Figure 8.2: O/C ratio of the protective leaf Pilot-scale plasma, $P_N = 200$ W, $p = 50$ Pa, $t = 0 - 120$ s, E-mode glow and H-mode afterglow treatments.

The seed coats of many plant species' seeds are also covered with cuticles (De Giorgi et al. 2015, e1005708), and a comparable effect of cuticular wax oxidation was found in quinoa seeds treated with air plasma, either as an atmospheric pressure surface DBD or low pressure capacitively coupled RF discharge (Gómez-Ramírez et al. 2017, 5924). Nasturtium seed treatment with an atmospheric pressure DBD in an air/helium mixture also increased O with treatment time, rapidly reaching saturation (Molina et al. 2018).

8.4.1.2 H-mode afterglow treatment

In Figure 8.2, we have already shown that the O/C ratio of the protective leaf surface increases more gradually during pilot-scale H-mode afterglow treatment compared to the E-mode glow treatment. When the same discharge parameters are used to produce both the E-mode glow and the H-mode afterglow plasma, the slope at which the O/C ratio increases in the afterglow treatment is approximately half the slope of the glow treatment. At different discharge parameters, namely higher P_N and lower working pressure, the increase is somewhat higher; the O/C ratio reaches a value of about 0.7, as opposed to about 0.5, within 60 s of treatment (Figure 8.3).

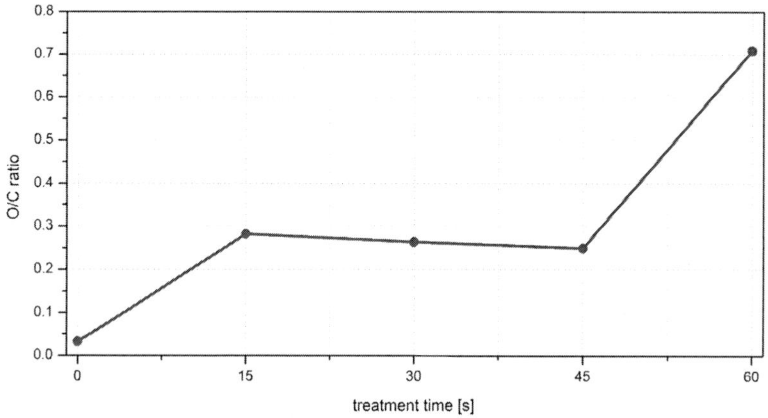

Figure 8.3: O/C ratio of unpeeled garlic, pilot-scale plasma AG. P_N = 300 W, p = 30 Pa, t = 0 – 60 s.

In a plasma system with a borosilicate reactor chamber, neutral oxygen atoms are reasonably long-lived species, as they do not readily recombine on the glass reactor walls due to the low recombination coefficient of glass (Wickramanayaka et al. 1991, 2897-2900). While n_O drops with increasing distance from the immediate glow discharge, it is nevertheless maintained at a considerable level throughout the discharge tube, assisted by the gas flow downstream of the discharge (Kutasi et al. 2014). In addition, in inductively coupled discharges, weak capacitive coupling is present throughout the reactor. The combination of these two effects exposes the sample in the afterglow treatment position to a lower, but still considerable flux of O atoms compared to that of the E-mode glow.

In the two experiments considering treatment in the pilot-scale reactor afterglow, the XPS analysis result at the only repeated treatment time (30 s) is comparable, despite the difference in generator power and working pressure (200 W and 50 Pa as opposed to 300 W and 30 Pa). This may indicate that treatment time is the most important functionalization parameter when treating the garlic clove with H-mode afterglow. In the same plasma system, it has previously been shown that after transition to the H-mode, increasing the generator power does not critically increase O atom density, especially at low pressures (Zaplotnik et al. 2011, 55001). While the considerable increase in the density of charged and highly excited species in the H-mode must be considered, the effect of these species is practically negligible outside of the immediate H-mode glow discharge.

8.4.1.3 Pulsed treatment

To avoid potential harmful effects of prolonged temperature increases on the sample, a pulsed treatment regimen may be applied. Although in the case of the E-mode discharge used here, temperature increases are not substantial, two pulsed regimens in the pilot-scale E-mode glow plasma were applied nonetheless: $P_N = 200$ W, $p = 70$ Pa, $t = 30$ s (6×5 s on, 5 s off), and $P_N = 300$ W, $p = 70$ Pa, $t = 60$ s (6×10 s on, 5 s off). Surface O/C ratio after such treatments is shown in Figure 8.4.

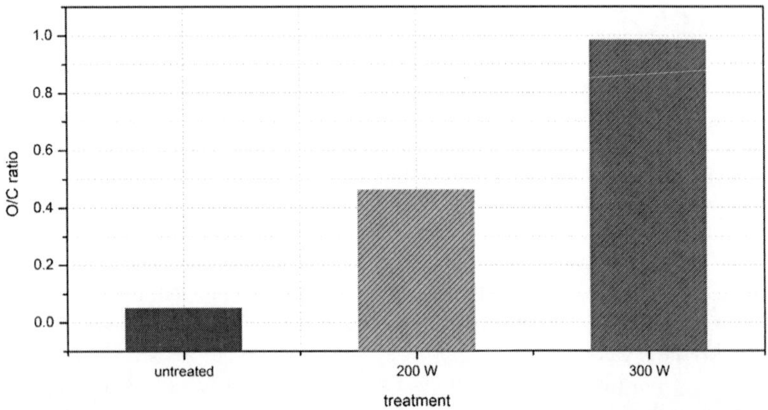

Figure 8.4: O/C ratio of unpeeled garlic, pilot-scale plasma, pulsed. $P_N = 200$ W, $p = 70$ Pa, $t = 30$ s (6×5 s on, 5 s off), and $P_N = 300$ W, $p = 70$ Pa, $t = 60$ s (6×10 s on, 5 s off).

Unsurprisingly, the O concentration and O/C ratio increases are higher in the treatment regimen with higher P_N and longer total treatment time. Comparing the O/C ratio between the 200 W treatments, pulsed (Figure 8.4) and continuous (Figure 8.2), it is evident that at the same treatment time (30 s), the O/C ratio is higher in the continuous treatment. This is due to the lower working pressure in that experiment (50 as opposed to 70 Pa), which produces a higher n_O (Figure 6.2). As will be shown below, the effect of the continuous and pulsed treatments on the sprout and root growth does not differ significantly if both treatments use the same discharge parameters.

8.4.2 Clove surface

8.4.2.1 Pilot-scale treatment

As with the protective leaf, oxygen plasma treatment also increases O concentration at the peeled clove surface (Figure 8.5). Compared to the protective leaf (Figure 8.1), the O increase at the peeled clove surface is more gradual at the same discharge parameters. The same is true for the O/C ratio, which within 120 s of treatment increases up to about 0.6 (Figure 8.6), as opposed to about 2.0 for the protective leaf (Figure 8.2).

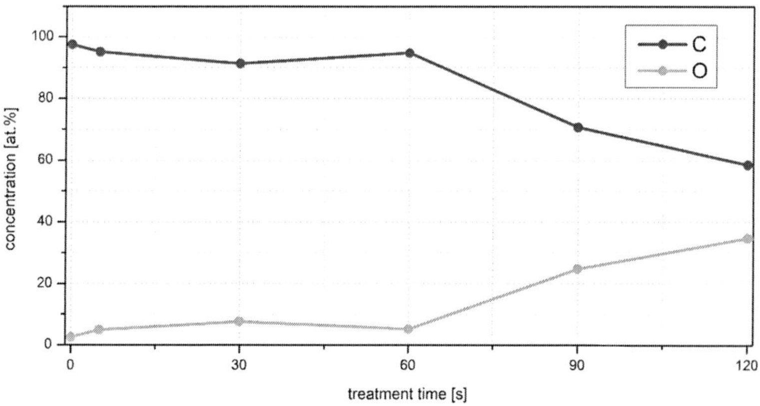

Figure 8.5: C and O concentration of the clove surface Pilot-scale E-mode glow, $P_N = 200$ W, $p = 50$ Pa, $t = 0 - 120$ s.

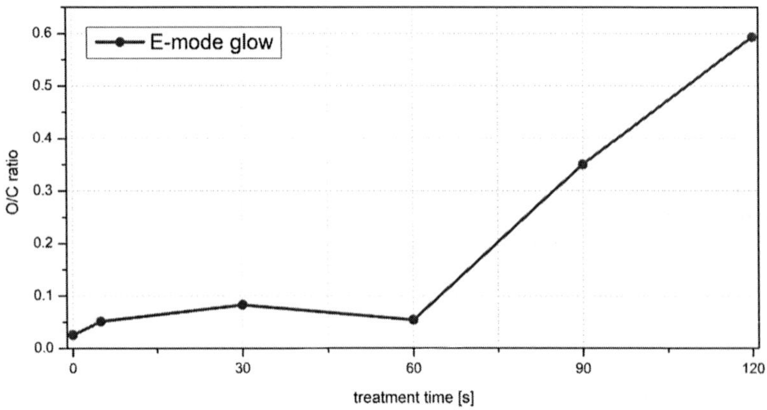

Figure 8.6: O/C ratio of the clove surface. Pilot-scale plasma, $P_N = 200$ W, $p = 50$ Pa, $t = 0 - 120$ s.

The more gradual changes at the clove surface, compared to the protective leaf, may be explained as follows. In the wax oxidation process, the most important initialization reaction is abstraction of a hydrogen atom from the alkyl chain. However, oxidation after abstraction of H from a terminal alcohol group is the most important mechanism for further oxidation (Dufour et al. 2014, 224007). Therefore, cuticle waxes already containing terminal –OH groups may undergo accelerated further oxidation, as well as the following etching.

8.4.2.2 Industrial-scale treatment

The oxidation seen at the peeled clove surface after industrial-scale plasma treatment is limited to a few at.%, despite higher P_N at the generator. This is also reflected in the O/C ratio (Figure 8.7). The slight changes follow no trend with changing treatment pressure. No elements other than C and O appear at the clove surface even after plasma treatment.

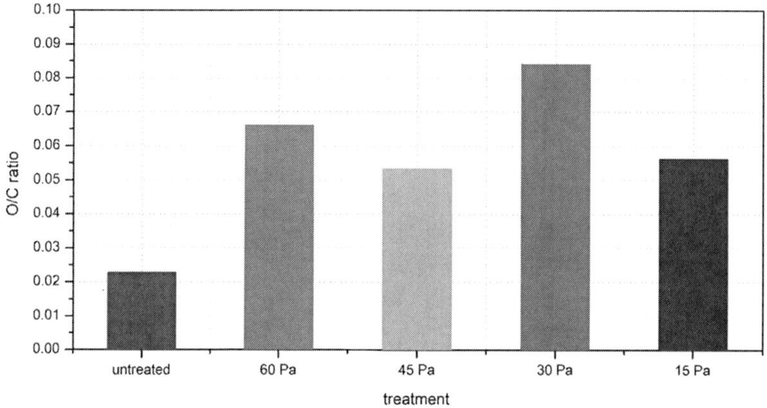

Figure 8.7: O/C ratio of peeled garlic, industrial-scale plasma. $P_N = \sim 700 - 900$ W, $p = 15 - 60$ Pa, $t = 60$ s.

While it is difficult to directly compare the treatments in the pilot-scale and industrial-scale plasma systems, we can ascribe this difference to the experimental conditions. Although the nominal power is about 4 × larger in the industrial-scale experiment (~ 800 W, as opposed to 200 W), the volume of the industrial-scale reactor is larger than that of the pilot-scale reactor by nearly a factor of 100 (about 63 L, as opposed to about 0.8 L). Even without an assumption of uniform plasma distribution throughout the discharge tube, it is clear that the plasma power density will be much larger in the more compact pilot-scale reactor. Therefore, the treatment in the industrial-scale plasma in this case is gentler. Further, the absence of mineral elements (silicon, calcium, potassium) indicates that no substantial etching of the surface takes place during plasma treatment.

8.4.2.3 Peeled clove surface under the treated protective leaf

As stated, purpose of the protective leaf is to protect the clove from unfavorable environmental factors. Therefore, it would be of interest to know to what degree plasma can affect the clove surface if the protective leaf is intact during plasma treatment. To this end, an unpeeled clove may be treated in the pilot-scale plasma, its protective leaf then removed, and the clove surface under it analyzed using XPS. Results of such an analysis show that oxygen concentration at the clove surface increases only slightly, about 12 × less than when a peeled clove is treated directly at the same discharge parameters. This indicates that when still properly attached, the protective leaf presents an effective barrier against plasma reactive species penetration to the clove surface.

8.5 Carbon chemical bonding

Additional data regarding surface chemistry can be obtained from high-energy resolution C 1s spectra. Differences in the area of fitted sub-peaks correspond to the relative frequency of various chemical bonds appearing on the surface. In the case of oxygenated waxy surfaces, four sub-peaks at the following binding energies should be considered for the main carbon peak:

- ~ 284.8 eV for C–C/C–H bonds;
- ~ 286.1 eV for C–O bonds;
- ~ 287.5 eV for C=O bonds; and
- ~ 288.9 eV for O–C=O bonds (Gerenser 1987, 303-318).

8.5.1 Protective leaf

Comparison of high-resolution C 1s spectra of the unpeeled clove (protective leaf) surface is shown in Figure 8.8. The area of sub-peaks corresponding to different chemical bonds is presented in Figure 8.9. The area of the main sub-peak at ~ 284.8 eV, corresponding to C–C/C–H bonds, rapidly diminishes with increasing treatment time, up to including 30 s of treatment. With longer treatment, the area of this peak gradually begins to increase again. Simultaneously, the area of the sub-peaks at ~ 286.1 eV for C–O bonds, ~ 287.5 eV for C=O bonds, and ~ 288.9 eV for O–C=O bonds increases within the first 15 – 60 seconds of treatment, and thereafter begins to gradually decrease again.

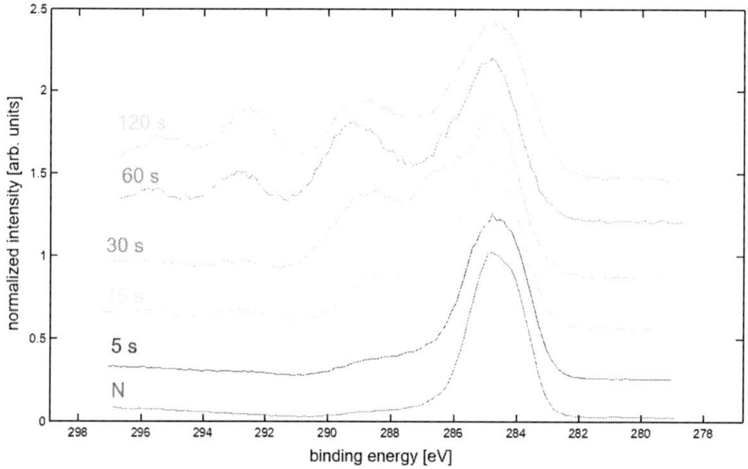

Figure 8.8: High-resolution C 1s spectra of unpeeled garlic. Pilot-scale E-mode glow, $P_N = 200$ W, $p = 50$ Pa, $t = 0 - 120$ s.

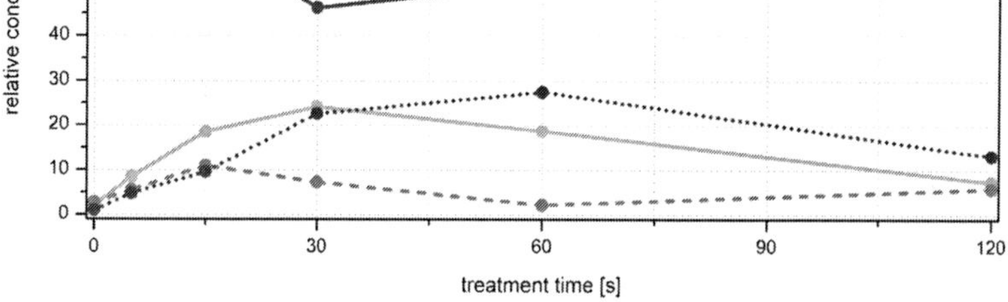

Figure 8.9: C 1s sub-peak area of unpeeled garlic. Pilot-scale E-mode glow, $P_N = 200$ W, $p = 50$ Pa, $t = 0 - 120$ s.

In the case of untreated protective leaf, the main peak, which corresponds to C–C/C–H bonds, takes up > 90 % of the total peak area. This is expected, as the cuticular waxes consist mostly of long hydrocarbon chains, where

C–C and C–H bonds are ubiquitous. The sub-peaks of C–O, C=O and O–C=O bonds appear at the higher binding energy side of the main peak. They represent the terminal oxygen-containing functional groups of the cuticular wax compounds, as well as the ester bonds of the wax compound dimers which form from the alcohol and carboxylic acid monomers (Riederer and Müller 2006). In comparison, a native, untreated cherry laurel leaf cuticle was analyzed to have relative peak areas of 87.1, 7.6, 3.4, and 1.9 %, for C–C/C–H, C–O, C=O, and O–C=O peaks, respectively (Perkins et al. 2005, 123-134). The higher relative presence of all oxygen-containing functional groups compared to garlic is due to the differences in cuticular wax composition between species and plant organs.

As treatment time progresses up to 30 s, the area of the main peak diminishes from about 94 % to about 46 % (Figure 8.9). This is in line with the rapid decrease of the surface C concentration in the same time frame (Figure 8.1). The concurrent increase in the area of the remaining sub-peaks corresponds to the O concentration increase. All of this indicates that functionalization of the protective leaf surface is taking place, as oxygen is introduced onto the surface through mechanisms further presented below. Further oxidation of already present functional groups also takes place. This evolution in the C 1s peak shape is typical for organic material exposure to oxidant plasmas (Gerenser 1987, 303-318), and was also recorded on agricultural seed surfaces. A low-pressure RF discharge and an atmospheric DBD, both in air, decreased the C–C/C–H peak and increased the presence of oxygen-containing functional groups on quinoa seeds (Gómez-Ramírez et al. 2017, 5924). On nasturtium seeds treated in air/He atmospheric DBD, increases in C=O and O–C=O also occurred, but the relative sub-peak areas reached saturation already after 30 s (Molina et al. 2018).

Out of the sub-peaks for carbon-oxygen bonds, the C–O peak increases the most rapidly. This represents the introduction of new –OH groups onto the alkyl chains, and is therefore consistent with the functionalization which is taking place. The C=O and O–C=O peaks increase comparably up to 15 s of treatment. Then, the C=O peak begins to decrease, and the O–C=O peak keeps increasing up to 60 s of treatment. As aldehydes are not particularly likely to form in the oxidation process, and are likely to be rapidly oxidized further to carboxylic acids (Dufour et al. 2014, 224007), it makes sense that the O–C=O peak is more prominent than the C=O peak at most of the treatment times.

Let us further consider the surface morphology (Figure 9.1). It is clear that up to 30 s of treatment, there is no intense etching, and the surface changes appear more in the form of reshaping of the existing morphology. However, at t = 30 s and onward, etching is clearly evident. As seen in

Figure 8.9, after 30 s, the C–C/C–H peak area increases again, while the remaining sub-peaks decrease. This is due to the intense chemical etching which begins around that time and exposes deeper, not yet oxidized layers. This equilibrium between functionalization and etching is reflected in the more gradual surface C content decrease and O content increase, as well as in the increasing relative area of the C–C/C–H component of the C 1s peak.

8.5.2 Clove surface

On the clove surface, within the first 60 s of treatment, the relative area of C 1s sub-peaks barely changes (Figure 8.11). The O/C ratio also remains relatively unchanged within this time frame (Figure 8.6). The C–O peak decreases, and the C=O and O–C=O peaks increase slightly, likely as oxidation of existing terminal –OH groups takes place. After 60 s, with the increase of O concentration and decrease of C concentration (Figure 8.5), the C–O peak increases, along with more gradual increases of the C=O and O–C=O peaks. Comparison of high-resolution C 1s spectra of the peeled clove surface is shown in Figure 8.10.

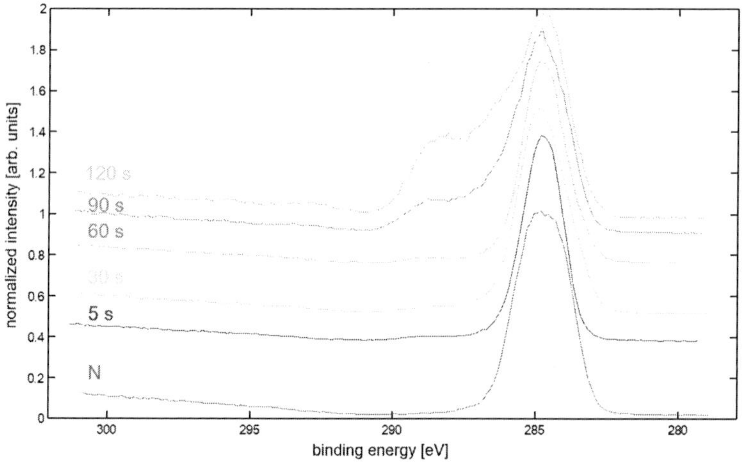

Figure 8.10: High-resolution C 1s spectra of peeled garlic. Pilot-scale E-mode glow, $P_N = 200$ W, $p = 50$ Pa, $t = 0 – 120$ s.

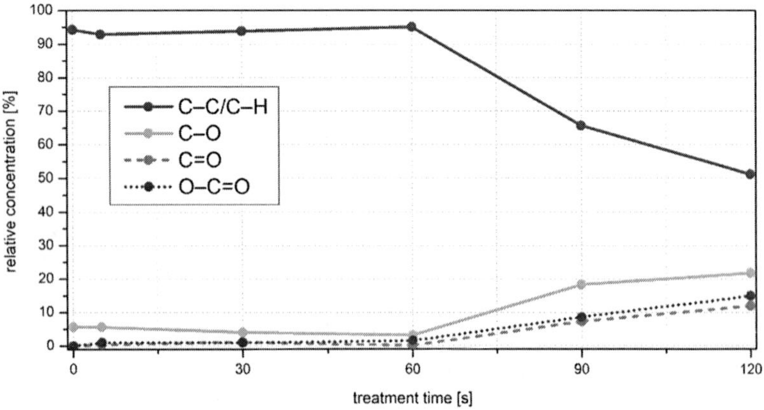

Figure 8.11: C 1s sub-peak area of peeled garlic. Pilot-scale E-mode glow, $P_N = 200$ W, $p = 50$ Pa, $t = 0 - 120$ s.

The behavior of curves for all sub-peaks between 60 and 120 s of treatment is very similar to what we see within the first 15 s of treatment of the unpeeled clove. This indicates that the same processes are taking place on the peeled clove surface. However, they start and progress more slowly.

Naturally, the initial chemical composition of the cuticle affects the behavior during plasma treatment. Cucumber and pepper seeds, which had a strongly oxygenated surface (18 – 21 at.% O) already when untreated, also showed relative C–O peak areas over 40 % after as little as 4 s of atmospheric air surface DBD treatment (Štěpánová et al. 2018, 1700076). As the peeled clove surface contains less oxygen than the protective leaf, it is expected that its oxidation will proceed more slowly.

8.5.3 Surface oxidation mechanisms

We have seen that during exposure to oxygen plasma, oxygen concentration at the garlic clove surface increases, and that the chemical functionality also changes with plasma exposure time. As with polymer materials, the hydrocarbon chains of the cuticle compounds are targets for functionalization by the reactive plasma species, particularly ROS in oxygen-containing discharges (Kitazaki et al. 2012, 01AE01). The neutral oxygen atom, O, is the most relevant oxidizing agent in oxygen plasma. The initial oxidation steps of cuticular wax components result in the introduction of hydroxyl and carbonyl groups into the hydrocarbon chain structure. The main products will be (secondary) alcohols, as well as their corresponding ketones. The alcohols can be further oxidized to carbonyl or carboxyl groups (Suhr 1983, 1-61; Patiño et al. 1995, 159-171). This is supported by the results of garlic cuticle analysis by XPS (Figure 8.9, Figure 8.11) as well as results of chemical characterization of tomato fruit cuticle by nuclear magnetic resonance (NMR) (Zhang et al. 2013, 425-432).

The oxidation of hydrocarbon chains is initiated by the formation of an alkyl radical, which occurs via a hydrogen abstraction by atomic oxygen:

$$\mathrm{RH} + \mathrm{O} \longrightarrow \mathrm{R}^{\bullet} + {}^{\bullet}\mathrm{OH}. \tag{8.1}$$

In linear hydrocarbon chains, such as those of cuticular waxes, this reaction would preferentially take place on secondary atoms (somewhere within the chain, as opposed to on either end) (Patiño et al. 1995, 159-171). Further, a cuticular wax reaction with an oxygen atom may also split an alkyl chain directly. However, the activation energy (E_A) for this reaction is higher than the E_A for hydrogen abstraction (Dufour et al. 2014, 224007).

To obtain an oxygen plasma, pure oxygen gas is introduced into the vacuum system. However, it should be noted that a significant amount of OH radicals is present in the system during plasma treatment, especially of peeled cloves (Figure 5.2). We can attribute this in part to water originating from the cloves during treatment. However, water vapor plasma is mildly oxidizing as well, and OH radicals may dissociate further to form O and H radicals (Gorjanc et al. 2009, 557-567). Further, in addition to those formed by dissociation of the water molecule, OH radicals form in oxygen plasma anyway through hydrogen abstraction. Therefore, the effect of oxygen plasma containing some water vapor will likely be similar to that of pure oxygen plasma, just somewhat milder. This may also contribute to the slower oxidation process of the peeled clove surface compared to the protective leaf surface.

8.6 Other elements

8.6.1 Origin of other elements

With longer plasma treatment times of the garlic cloves, XPS begins to detect elements other than carbon and oxygen at the protective leaf and clove surfaces. As the cuticle is known to consist simply of oxygenated hydrocarbons, it is interesting to examine which additional elements appear, and what their origin could be in the context of biological components of the garlic clove as well as the plasma chemistry.

The etching effect, and thereby the exposure of deeper layers of the protective leaf, is responsible for the appearance of some of the elements. When examining surface morphology, it will be clearly seen that with longer plasma exposure time, etching gradually uncovers deeper layers of the protective leaf. Thereby, elements or compounds not found at the cuticle surface may also be uncovered. The reshaping of deeper layers of the storage leaf surface appears to begin only with even longer treatment times, which is reflected in the time of appearance of additional elements. Some of the differences between the behavior of the protective leaf and storage leaf surfaces may also reflect differences in cuticle composition between the two organs.

8.6.2 Protective leaf

In the case of the unpeeled cloves treated in the pilot-scale E-mode glow plasma (Table 8.2, Figure 8.12), calcium, sulfur and silicon concentration at the treated protective leaf surface increase. Magnesium and potassium also appear in concentrations under 1 at.%, sodium is detectable for treatment times of 5 and 15 s, and chlorine for treatment times of 0 and 15 s. Interestingly, nitrogen is present on untreated and briefly treated surfaces, but gradually disappears with longer treatment.

Table 8.2: Surface analysis of unpeeled cloves, pilot-scale plasma. Elemental composition [at.%]. $P_N = 200$ W, $p = 50$ Pa, $t = 0 - 120$ s, glow and afterglow (AG) treatments.

t [s]	C	O	Ca	S	Si	Mg	N	Cl	K	Na
0	90.2	8.2	0.1	/	/	/	1.4	0.1	/	/
5	77.5	20.0	0.2	0.2	/	/	1.7	/	/	0.4
15	65.3	30.1	1.6	0.5	0.4	/	1.1	0.2	/	0.9
30	44.2	46.0	6.8	1.5	0.7	0.4	0.3	/	/	/
60	38.0	48.6	9.0	2.2	0.6	0.8	0.3	/	0.4	/
120	27.6	57.2	8.6	4.4	1.0	0.6	/	/	0.7	/
AG 30	79.1	18.9	/	/	/	/	1.3	0.1	/	0.6
AG 120	46.9	43.6	4.3	1.6	0.3	0.5	0.8	0.2	0.8	1.1

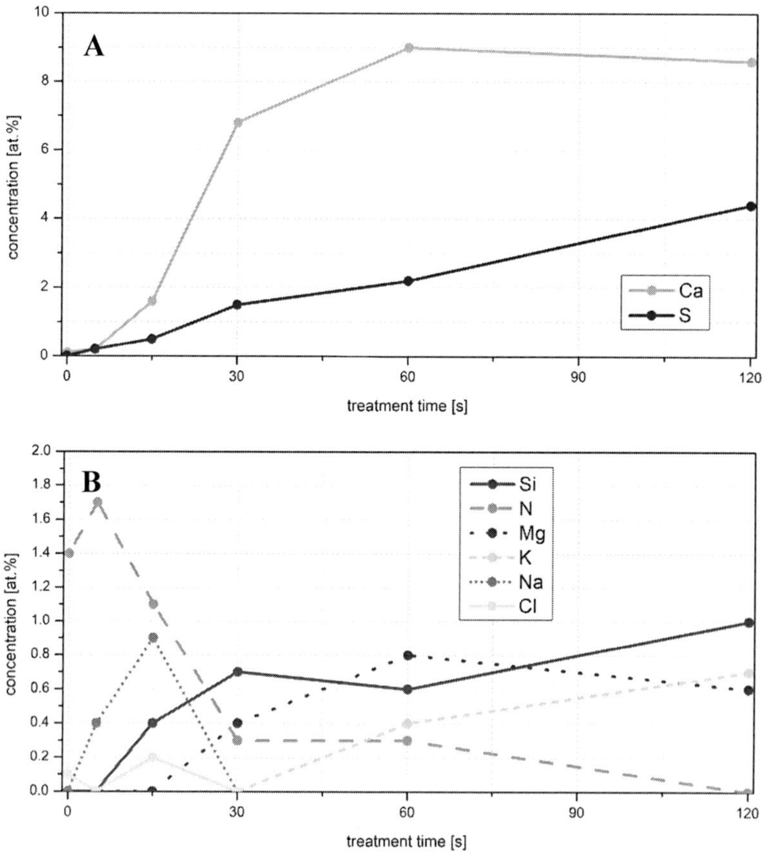

Figure 8.12: Minor elements concentration at the protective leaf surface. Pilot-scale E-mode glow, P_N = 200 W, p = 50 Pa, t = 0 – 120 s. (A) Calcium and sulfur. (B) Other elements.

After treatment with the H-mode afterglow, the surface elemental composition differs from the one seen after E-mode glow treatment (Table 8.3). This is a consequence of the difference in the plasma reactive species composition between the two treatment conditions.

Table 8.3: Surface of unpeeled cloves, pilot-scale H-mode AG. Elemental composition [at.%]. $P_N = 300$ W, $p = 30$ Pa, $t = 0 – 60$ s.

t [s]	C	O	Ca	Si	N	K
0	96.9	3.1	/	/	/	/
15	75.8	21.4	0.6	1.5	0.7	/
30	77.4	20.4	0.5	0.9	0.8	/
45	77.9	19.4	/	0.5	2.3	/
60	56.0	39.7	1.2	0.6	1.9	0.6

After plasma treatment with a pulsed regimen, a variety of elements appears at the surface prominently, in particular after the 300 W regimen (Table 8.4).

Table 8.4: Surface analysis of unpeeled cloves, pulsed treatment. Elemental composition [at.%]. Pilot-scale glow plasma, pulsed treatment regimens: $P_N = 200$ W, $p = 70$ Pa, $t = 30$ s (6 × 5 s on, 5 s off), and $P_N = 300$ W, $p = 70$ Pa, $t = 60$ s (6 × 10 s on, 5 s off).

treatment	C	O	Ca	S	Si	Mg	N	K	F
untreated	93.9	4.7	/	/	0.8	/	0.6	/	/
200 W, 30 s	63.5	29.4	3.9	0.9	0.5	/	1.1	0.6	/
300 W, 60 s	44.3	43.6	4.2	2.0	1.4	1.6	0.7	1.3	1.0

8.6.3 Clove surface

At the clove surface (Table 8.5, Figure 8.13), no element other than carbon and oxygen appears earlier than after 30 s of pilot-scale E-mode glow plasma treatment. Calcium, sulfur and silicon concentration all begin to increase later and reach lower values within 120 s of treatment than at the protective leaf surface. Magnesium only appears after 120 s, but nitrogen begins to increase after as little as 30 s. Fluorine also appears from 30 s onward.

Table 8.5: Surface analysis of peeled cloves, pilot-scale plasma. Elemental composition [at.%]. $P_N = 200$ W, $p = 50$ Pa, $t = 0 - 120$ s.

t [s]	C	O	Ca	S	Si	Mg	N	P	F
0	97.5	2.5	/	/	/	/	/	/	/
5	95.1	4.9	/	/	/	/	/	/	/
30	91.3	7.6	/	/	0.3	/	0.3	/	0.5
60	94.8	5.2	/	/	/	/	/	/	/
90	70.7	24.8	1.5	/	0.5	/	2	/	0.3
120	58.5	34.7	2.5	0.1	0.6	0.4	2.5	/	0.7

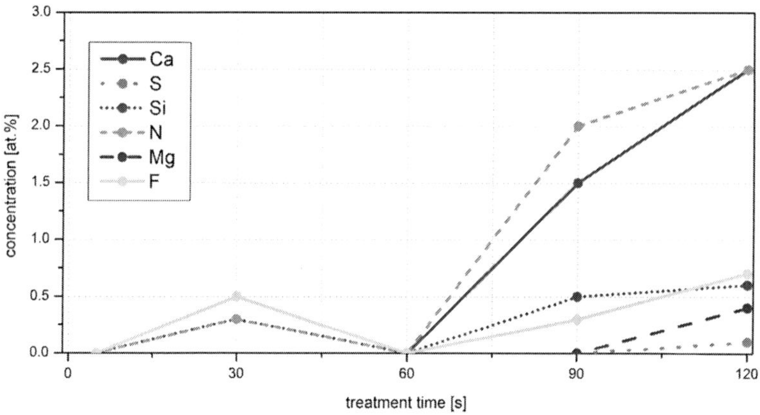

Figure 8.13: Minor elements concentration at the peeled clove surface. Pilot-scale E-mode glow, $P_N = 200$ W, $p = 50$ Pa, $t = 0 - 120$ s.

At the surface of peeled cloves treated in the industrial-scale plasma system, only carbon and oxygen appear, regardless of changes in working pressure (Table 8.6). As already mentioned, this is due to the gentler treatment, which results from the lower plasma power density in this large-scale reactor.

Table 8.6: Surface analysis of peeled garlic, industrial-scale plasma. elemental composition [at.%]. $P_N = \sim 700 - 900$ W, $p = 15 - 60$ Pa, $t = 60$ s.

treatment	C	O
N	97.8	2.2
60 Pa	93.8	6.2
45 Pa	94.9	5.1
30 Pa	92.2	7.8
15 Pa	94.7	5.3

Next, let us examine the possible origins of the elements detected at the plasma-treated garlic surfaces.

8.6.4. Sulfur

While the main constituents of cuticular wax, as well as the cutin polymer scaffold, only consist of carbon, oxygen and hydrogen, it is generally possible to find within the cuticle compounds such as alkaloids, which include heteroatoms in their structure (Riederer and Müller 2006). The majority of sulfur in garlic is found in non-protein amino acids, including volatile flavor compound precursors (Kamenetsky 2007, 123-172). Therefore, it is likely that the detected sulfur originates from allicin or other garlic thiosulfinates. The allicin precursor, alliin, is primarily produced in garlic leaves and translocated to other parts of the plant (Nasim et al. 2010, 59-63), where it is typically located inside of cells. After damage to the plasma membrane, an enzymatic reaction converts alliin to allicin, which gives garlic its pungent smell and taste (Bocchini et al. 2001, 37-43). It was shown that allicin does not permeate cuticle waxes efficiently when sprayed onto plant surfaces (Leontiev et al. 2018, 6763), and its precursor compound is even more polar. However, as we will show during the discussion of ATR-FTIR analyses below, it appears that plasma treatment at the selected discharge parameters does not affect the garlic surfaces deeply enough to affect the epidermal cells beneath the cuticle. Further, sulfur-containing compounds are mainly found in the storage leaf, while in the protective leaf, their concentration is much lower (Freeman 1975, 471-481). Nonetheless, if the plasma etching of the garlic surfaces is deep enough to release any of the sulfur-containing garlic compounds to the surface, or if they are normally found in the cuticle in small amounts, this may account for S appearance in the XPS spectra. Another possible origin of S in biological samples would be proteins, but a protein-related increase in S would need to be accompanied by an increase in N. However, in the case of garlic protective leaf, N concentration at the surface clearly decreases with increasing treatment time.

Sulfur only appears in the XPS profile of peeled cloves after 90 s of treatment, and in very low amounts. This is surprising, as the concentration of thiosulfinates in the fresh storage leaf is around 700 × higher than in the fresh protective leaf (Freeman 1975, 471-481). While most of these compounds are found in cells in the bulk of the storage leaf, the only plausible explanation for such a late appearance of sulfur during peeled clove treatment is the milder treatment due to increased water presence in the system, and thereby, the slower chemical etching process.

8.6.5. Calcium

On the unpeeled clove surface, the most prominent element concentration increase with treatment time is calcium, up to about 9 at.% (Figure 8.12). Ca^{2+} ions are capable of penetrating the cuticle despite their polar nature, likely at the site of stomata, or aqueous pores in the cuticle (Schreiber 2005, 1069-73). However, on the garlic protective leaf surface, stomata are practically missing (Mann 1952, 195-251), and it is therefore unlikely that the calcium originates from within epidermal cells. On the other hand, Ca^{2+} ions are bound by free carboxyl groups of the pectin in the layer which connects the cuticle and the plant cell wall (Hepler 2005, 2142-55). If the pectin layer is at least partly exposed during plasma treatment, this would explain the increase in calcium concentration at the surface.

Similar as with sulfur, calcium content also begins to increase with peeled clove treatment, but is detected only after longer treatment times, and in lesser amounts (Figure 8.13).

8.6.6. Silicon

The origin of the detected silicon is not completely clear. In polymers, silicon may be detected at the surface after plasma treatment, as it is an additive in polymer production or processing, and is then exposed during plasma etching of the polymer (Vukušić et al. 2018). However, the garlic used here is unprocessed, lacking any additives. Although XPS analysis of processed garlic peel did not detect any silicon (Liu et al. 2014, 2054-2063), it is known that certain plants may accumulate silicon in a layer underneath the cuticle, possibly due to its beneficial mechanical effects in protection from plant stress (Ma et al. 2001, 17-39). This accumulation, and the subsequent exposure of silicon with progressing plasma treatment time, seems like the most likely origin for its detection.

Similar as with sulfur and calcium, silicon concentration also begins to increase with peeled clove treatment, but is detected only after longer treatment times, and in lesser amounts (Figure 8.13).

8.6.7 Nitrogen

Interestingly, nitrogen concentration at the protective leaf surface starts at ~ 1.5 at.% and decreases with plasma treatment time, until none is detected at 120 s of treatment (Figure 8.12). The most obvious source of nitrogen in living matter is protein. Protein content of the entire garlic protective leaf is about 8 % (Abdel-Fattah and Edrees 1972, 871-877), but the protein is found as cellular components, and therefore, in deeper layers of the protective leaf than those accessed by etching. None of the cuticular waxes, polymers or other outer layer components contain any nitrogen in their structure. It is therefore reasonable to suggest that the source of nitrogen is external, for example, that nitrogen is present on the protective leaf surface due to fertilization of garlic with nitrogen-containing fertilizer.

This suggestion is supported by the fact that on the peeled clove surface, which is tightly wrapped from the outside by the protective leaf, no nitrogen is detected initially, although it does appear at the surface after 90 s of treatment and onwards (Figure 8.13). This later appearance of nitrogen on the peeled clove surface may be due to the presence of enzymes in the cell wall, which in opposition to the protective leaf epidermal cell wall, is not sclerified (Mann 1952, 195-251).

8.6.8 Magnesium and potassium

Magnesium and potassium also increase on garlic surfaces after plasma treatment. As suggested for plasma-treated quinoa seeds, these elements likely diffuse toward the treated surface as a consequence of plasma etching (Gómez-Ramírez et al. 2017, 5924). Indeed, they only appear in XPS spectra at treatment times where etching of the protective leaf is already apparent in the morphology as observed by SEM (Figure 8.12). At the peeled clove surface (Figure 8.13), magnesium only appears after 120 s of treatment, and potassium does not appear at all.

8.6.9 Sodium and chlorine

On the protective leaf surface (Figure 8.12), the random appearance of sodium and chlorine hints at contamination. Neither element appears at the peeled clove surface.

8.6.10 Fluorine

On the clove surface, from 30 s of treatment onward, minor amounts of fluorine appear (Figure 8.13). To the best of our knowledge, fluorine has no physiological function in garlic, but may be accumulated in plants from the environment, either from pollution or compounds such as pesticides or fertilizers. It is also possible that fluorine contamination originates from the plasma system.

8.7 Infrared spectroscopy

8.7.1 Untreated surfaces

In addition to high-energy resolution XPS spectra, more information regarding surface chemical bonding is found in ATR-FTIR spectra of unpeeled and peeled garlic surfaces. To understand this, we must first look at the spectra recorded for untreated surfaces. Figure 8.14 shows a comparison of ATR-FTIR spectra of the untreated protective leaf and peeled clove. Differences and similarities are apparent.

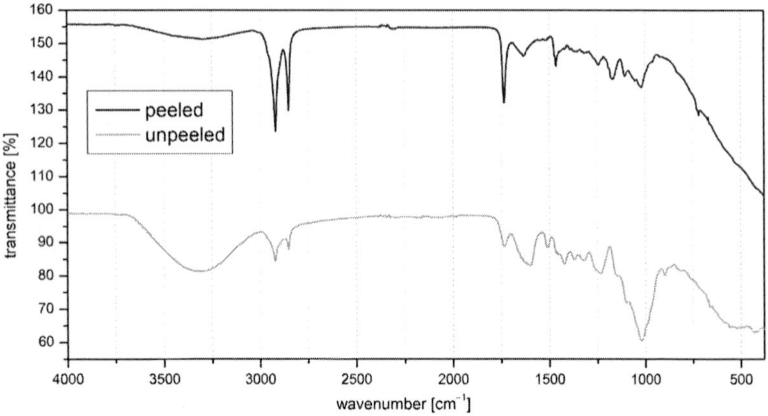

Figure 8.14: ATR-FTIR spectra of untreated clove surfaces.

The differences between the two spectra emphasize the difference in their chemical composition and correspond well to XPS results of the untreated surfaces (Table 8.2). The higher oxygen content of the protective leaf is reflected here in the prominent O–H peak at ~ 3300 cm^{-1}, which corresponds to the alcohol O–H stretching and indicates the presence of more –OH functional groups in the waxy compounds of the cuticle. Conversely, the sharp, tall peaks of C–H bond stretching (~ 2850 and 2920 cm^{-1}), much more prominent in the peeled clove spectrum, correspond to the high alkane chain content of that surface, which is seen in XPS results as a high concentration of carbon. The peak at 1465 cm^{-1}, seen only in the peeled clove spectrum, further stands for C–H bending of methylene groups in alkanes (LibreTexts 2019).

The major differences between the two spectra start at ~ 1750 cm^{-1} and extend into the fingerprint region. The peak at ~ 1732 cm^{-1} represents the C=O bond (LibreTexts 2019; Zahoranová et al. 2018, 969-988). It is taller and thinner at the peeled clove surface, indicating a stronger presence of the C=O bond, either as an aldehyde/ketone, or as part of a carboxylic acid or ester. As the XPS results indicate less initial oxygen presence at the peeled clove surface (Table 8.2), it is likely that C=O bond signal here derives from groups present in deeper layers of the clove surface, possibly the ester bonds of the cutin polymer scaffold.

In addition to the O–H stretching peak at ~ 3300 cm^{-1}, both untreated spectra exhibit a peak at ~ 1632 cm^{-1}, indicating O–H absorption (Hardin et al. 2013, 331-339). On the peeled clove surface, this peak is clearly distinct, but slightly smaller. On the protective leaf surface, it is taller, but hidden as a shoulder within another peak. This peak, found at ~ 1600 cm^{-1} only in the unpeeled clove spectrum, appears to represent C=C stretching. It is complemented by the peak at 897 cm^{-1} for C=C bending (LibreTexts 2019; Ji et al. 2018, 145205). Unsaturated analogues of the typically saturated compounds may be present in cuticular waxes (Riederer and Müller 2006), and are a possible source of the C=C signal.

The differences in the peak intensities for O–H, C=O and C=C groups, as well as almost fully distinct fingerprint regions, hint at differences in the chemical composition of untreated protective leaf and storage leaf surfaces analyzed by FTIR. They could either be the consequence of different cuticular wax composition, or stem from a difference in cuticular wax thickness, which influences the depth of further layers included in the FTIR analysis, and thus the obtained spectrum. As cuticular wax composition within the same plant species is somewhat consistent, the second option seems more likely. We know that, as the clove is already protected by the protective leaf, the epidermis of the storage leaf is thin and delicate (Mann 1952, 195-251). It makes sense, then, that the cuticle covering it is also thinner, and that the IR beam during FTIR analysis penetrates into deeper layers, resulting in a difference in the fingerprint region of the FTIR spectrum.

8.7.2 Protective leaf

With plasma treatment, changes begin to appear in the infrared spectra of garlic surfaces. Figure 8.15 shows a comparison of ATR-FTIR spectra for untreated and treated surfaces of the protective leaf. Figure 8.16 compares the two spectra at the longest treatment times directly.

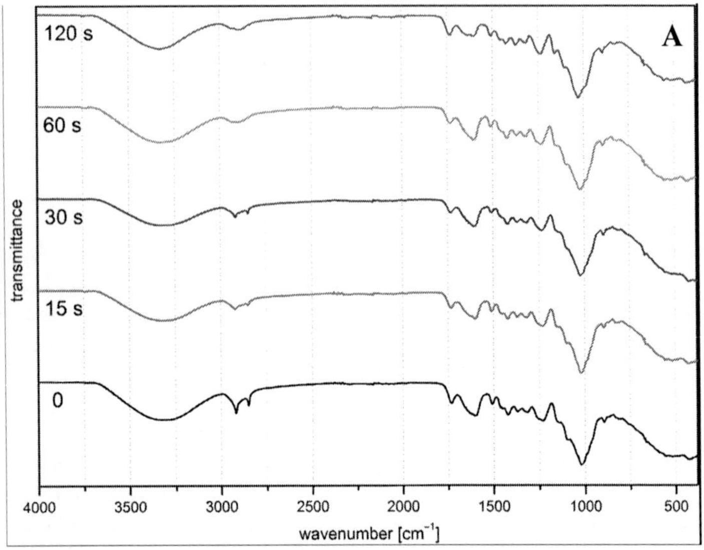

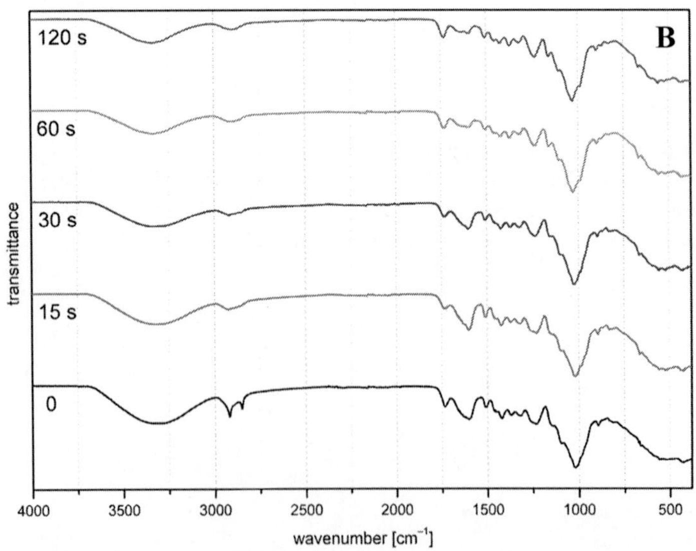

Figure 8.15: ATR-FTIR spectra, unpeeled clove. Pilot-scale E-mode glow, (A) $P_N = 200$ W, (B) $P_N = 300$ W, $p = 50$ Pa, t = 0 – 120 s.

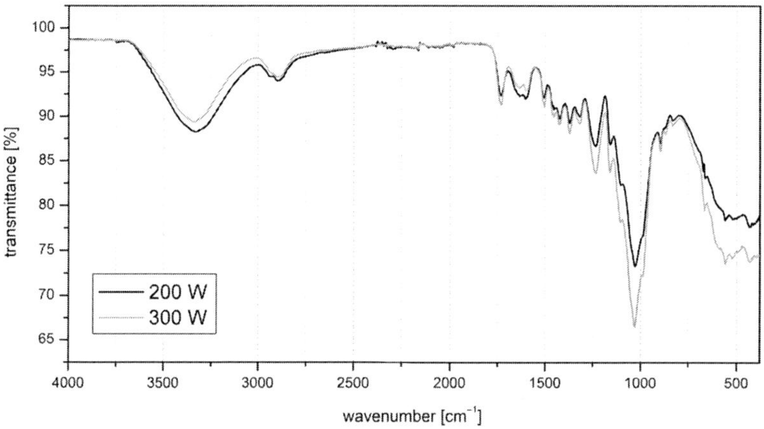

Figure 8.16: ATR-FTIR spectra, unpeeled clove, direct comparison. Pilot-scale E-mode glow, $P_N = 200 – 300$ W, $p = 50$ Pa, $t = 120$ s.

At both treatment powers used here (200 and 300 W), the intensity of C–H peaks at ~ 2920 and 2850 cm^{-1} diminishes with increasing treatment time. The sharp peaks become undistinguishable from the broader peak below; this occurs after 60 s for the lower treatment power (200 W), but already after 15 s for the higher treatment power (300 W). At the same time, the intensity of the O–H peak at ~ 3300 cm^{-1} also lowers slowly. From XPS results (Figure 8.1), we know that surface oxidation occurs, and O-containing groups replace hydrogen at the methylene chain surface, leading to C–H peak decrease. While new –OH groups are introduced, we also still see an O–H peak decrease, as these groups are also rapidly oxidized further.

Another feature of the spectrum that decreases in intensity is the C=C peak at ~ 1600 cm^{-1}, particularly at the longest treatment time of 120 s. It most likely represents the unsaturated C=C bonds present in the cuticular waxes, the composition of which may change with cuticle depth (Riederer and Müller 2006). The intensity loss is due to preferential oxidation of cuticular wax compounds at any available double bonds.

As shown in Figure 8.16, the spectra at 200 and 300 W are virtually identical even at the longest treatment time of 120 s. Interestingly, there are no changes to the 1800 – 1650 cm^{-1} region, where C=O stretching peaks typically appear. The C=O peak at ~ 1732 cm^{-1} remains comparable throughout the treatment. In addition, no changes at all are seen in the fingerprint region of the spectra, between 1500 and 400 cm^{-1}. This is due to the analysis depth of the method. ATR-FTIR is a less surface-sensitive technique than XPS. Its analysis depth, which depends on the depth of IR beam penetration, is typically between 0.5 and 2 μm (Mirabella 1993). Thus, in addition to the cuticle with its outermost, plasma-affected layer, the analyzed volume includes deeper layers of the protective leaf or clove (Taiz et al. 2014). The plasma functionalization processes affect only one atomic layer at a time, which are then gradually removed by chemical etching. Therefore, the depth in which chemical changes occur is very thin compared to the thickness of the layer analyzed by FTIR. If the composition of the cuticular waxes is fairly uniform along the cuticle depth, then these oxidative changes at the very surface do not detectably affect the fingerprint region. In available literature, this was also suggested as the reason why for radish seeds after low pressure capacitively coupled RF oxygen plasma treatment, there were no changes in ATR-FTIR spectra (Kitazaki et al. 2012, 01AE01). Similarly, there were no changes for nasturtium seeds treated in atmospheric pressure DBD in an air/He mixture with increasing treatment times, with the exception of a decrease in the O–H peak, which could be related to loss of water (Molina et al. 2018).

In addition, this result confirms that the plasma treatment within the selected conditions does not etch the cuticle away completely, or damage the epidermal cells. If that was the case, the fingerprint region of the spectrum would noticeably change, as the IR beam would penetrate deeper through a different set of compounds contained in the cells.

8.7.3 Clove surface

Changes due to plasma treatment are also reflected in infrared spectra of the peeled clove surface.

Figure 8.17 shows a comparison of ATR-FTIR spectra for untreated and treated surfaces of the peeled clove. The differences in the spectra at 200 and 300 W for a treatment time of 120 s are further shown in Figure 8.18.

At both treatment powers, the initially more prominent C–H stretching peaks at ~ 2920 and 2850 cm^{-1} decrease more slowly than at the unpeeled clove surface, indicating a slower surface oxidation process. The C–H bending peak decreases as well. As the plasma introduces new –OH groups to the surface, the O–H stretching and absorption peaks gradually increase. As above, the changes are more pronounced at longer treatment times, and occur earlier during treatment with higher P_N.

A sharp peak at ~ 1730 cm^{-1} falls with increasing treatment time, while a slightly broader peak at ~ 1630 cm^{-1} rises. The first belongs to C=O bonds, while the second may belong to the alkene C=C bond.

The fingerprint region appears comparable between untreated and plasma-treated clove surface for short treatment times. The remaining changes to the spectra occur mainly at the longest treatment times (120 s at P_N = 200 W, and 90 – 120 s at P_N = 300 W), when reshaping of the fingerprint region begins. As we will show below, these are treatment times that begin to negatively affect the sprout growth (Figure 11.9). Therefore, it is plausible that these treatment times begin to affect the epidermal cells, and damage the clove, or increase its vulnerability.

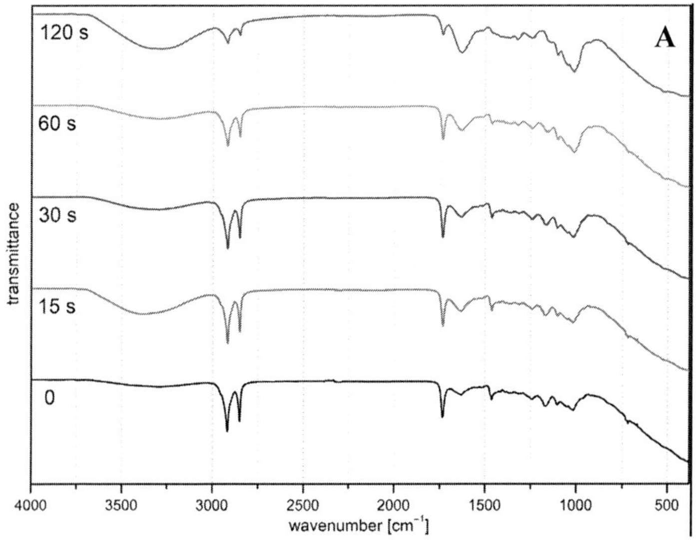

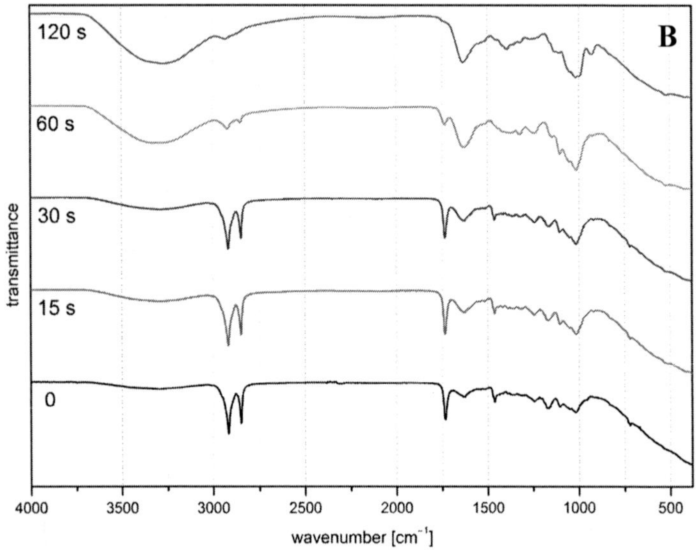

Figure 8.17: ATR-FTIR spectra, peeled clove. Pilot-scale E-mode glow, (A) $P_N = 200$ W, (B) $P_N = 300$ W, $p = 50$ Pa, $t = 0 – 120$ s.

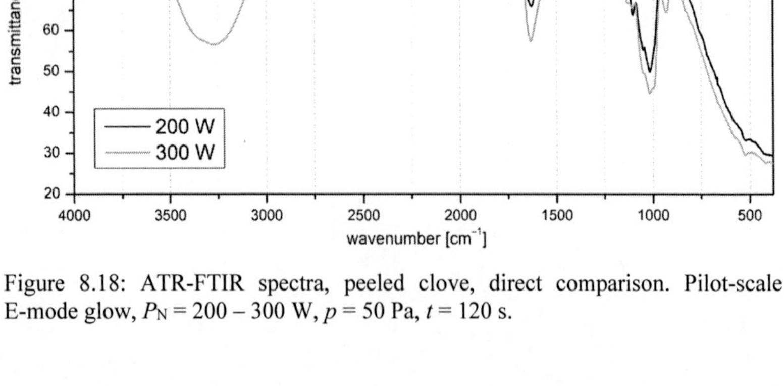

Figure 8.18: ATR-FTIR spectra, peeled clove, direct comparison. Pilot-scale E-mode glow, $P_N = 200 – 300$ W, $p = 50$ Pa, $t = 120$ s.

9. SURFACE MORPHOLOGY

9.1 Plasma effects on surface morphology

In addition to chemistry, morphology of the treated surfaces is affected by plasma, specifically by the process of etching. This occurs by two major mechanisms: physical sputtering and chemical etching (Fridman 2008). Energetic, positively charged ions are responsible for physical sputtering of the surface (Dhayal et al. 2006, 499-506), as they impinge on the polymeric surface with high kinetic energy, removing atoms by bond breakage in the process (Fridman 2008). On the other hand, the same oxidation mechanisms responsible for functionalization result in chemical etching with prolonged exposure. Namely, oxidative degradation of organic materials occurs when oxygen-containing functional groups are further oxidized and eventually decompose to form volatile compounds such as CO_2 and H_2O, which then enter the gaseous phase and are removed from the surface (Egitto et al. 1990, 321-422). Plasma etching increases the surface roughness, resulting in an increased surface to volume ratio (Phan et al. 2017, 417). A combined effect of both functionalization and etching is an increase in the surface energy of the treated material, which renders the surface more hydrophilic (Fridman 2008). The ratio of functionalization to etching depends on the reactive species present in the discharge, and therefore on the characteristics of the selected plasma as well as sample position relative to the discharge.

9.2 Methods of analysis

Surface morphology may be visualized either by obtaining microscope images or by directly characterizing the topography. Scanning electron microscopy (SEM) allows us to obtain images of material surfaces at very high magnifications. For garlic cloves, an organic sample material, it is possible to achieve clear images at magnifications up to at least 30,000 ×. The microscope fires a focused electron beam at the sample surface in a raster scan pattern. Signals emitted from sample surface atoms, such as secondary electrons, give information about sample topography. Before analysis, a thin layer of conductive coating must also be applied to make the otherwise organic surface conductive, a SEM requirement. In addition, as with surface chemistry analyses, conventional SEM devices also require placing the sample in high vacuum conditions. This again requires unpeeled and peeled garlic cloves to be treated differently. To analyze unpeeled cloves, the protective leaf may be removed after treatment, and is dry enough to be used directly. To analyze peeled cloves, pieces excised from the clove surface need to be freeze dried before coating and analysis.

Atomic force microscopy (AFM) provides a tool to directly characterize the topology of the samples. It uses a mechanical probe to raster scan an area of the sample surface and provide information about its topography based on the interactions between the tip of the probe and the sample. Since the probe can traverse the sample surface in atmospheric conditions, even organic samples such as garlic cloves may be analyzed directly, without further sample preparation.

9.3 Protective leaf

9.3.1 Pilot-scale glow plasma

It is well known that oxidative plasma etches polymer surfaces (Gorjanc and Mozetič 2014). This is also true for the waxy compounds that compose plant cuticles. Low pressure RF oxygen plasma has etched cuticular waxes on lamb's lettuce leaves (Grzegorzewski et al. 2010, 1145-1152) and tomato fruits (Zhang et al. 2013, 425-432). In plasma treatment of agricultural seeds, etching or erosion of the seed surface was seen after atmospheric pressure DBD treatment of wheat kernels in oxygen, air, argon, or nitrogen plasma (Meng et al. 2017, 1105-1119), or cotton seeds in air or nitrogen plasma (Wang et al. 2017, 5601), as well as atmospheric pressure micro DBD treatment of coriander seeds in nitrogen plasma (Ji et al. 2015, 1164-1173), or spinach seeds in air or nitrogen plasma (Ji et al. 2016, 117-28). Etching intensity often increases with treatment time (Ji et al. 2015, 1164-1173; Ji et al. 2016, 117-28; Meng et al. 2017, 1105-1119; Stolárik et al. 2015, 659-676). However, a quantification of the etching rate, or the thickness of the cuticle layer eroded during plasma treatment, is currently lacking in literature.

In the case of unpeeled garlic cloves, let us first examine the native protective leaf surface morphology. As seen in Figure 9.1, at low magnification (2,000 ×), its morphology consists of parallel grooves that run along the surface. The width of the grooves is about 15 – 20 μm, which suggests that they are the outlines of the outer epidermis cells. These are known to be about 500 μm long and run over the protective leaf surface like long parallel stripes. Their cell walls are lignified, and they are covered by a thin, smooth cuticle (Mann 1952, 195-251). This is seen at larger magnification (10,000 ×) of the untreated protective leaf as a wrinkled, but reasonably smooth surface.

Increasing surface roughness of the protective leaf becomes apparent with increasing plasma treatment time and/or generator power. The effect of pilot-scale E-mode glow plasma with increased exposure time is shown in Figure 9.1. After as little as 5 s of treatment, increased roughness is apparent in the form of surface bumps, which become more prominent after 15 s of treatment. The bumps are visually reminiscent of the roughness increase seen after oxygen plasma etching of polyethylene (Lehocký et al. 2003, 125-131), a polymer chemically related to cuticular waxes, as it is composed of long methylene chains. From 30 s onward, extensive etching is apparent. As the wax is etched away, the surface becomes nanostructured, and an irregular, ridge- or scaffold-like structure remains at the surface. The effect is increased with increasing the treatment time, and thereby, with the plasma reactive species dose received by the clove. This is seen as further changes to the structure roughness and compactness with increased treatment time.

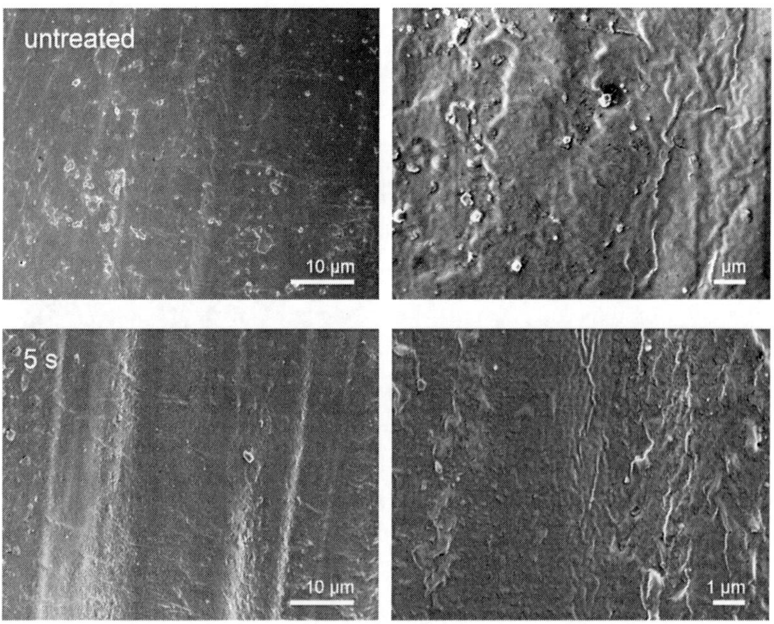

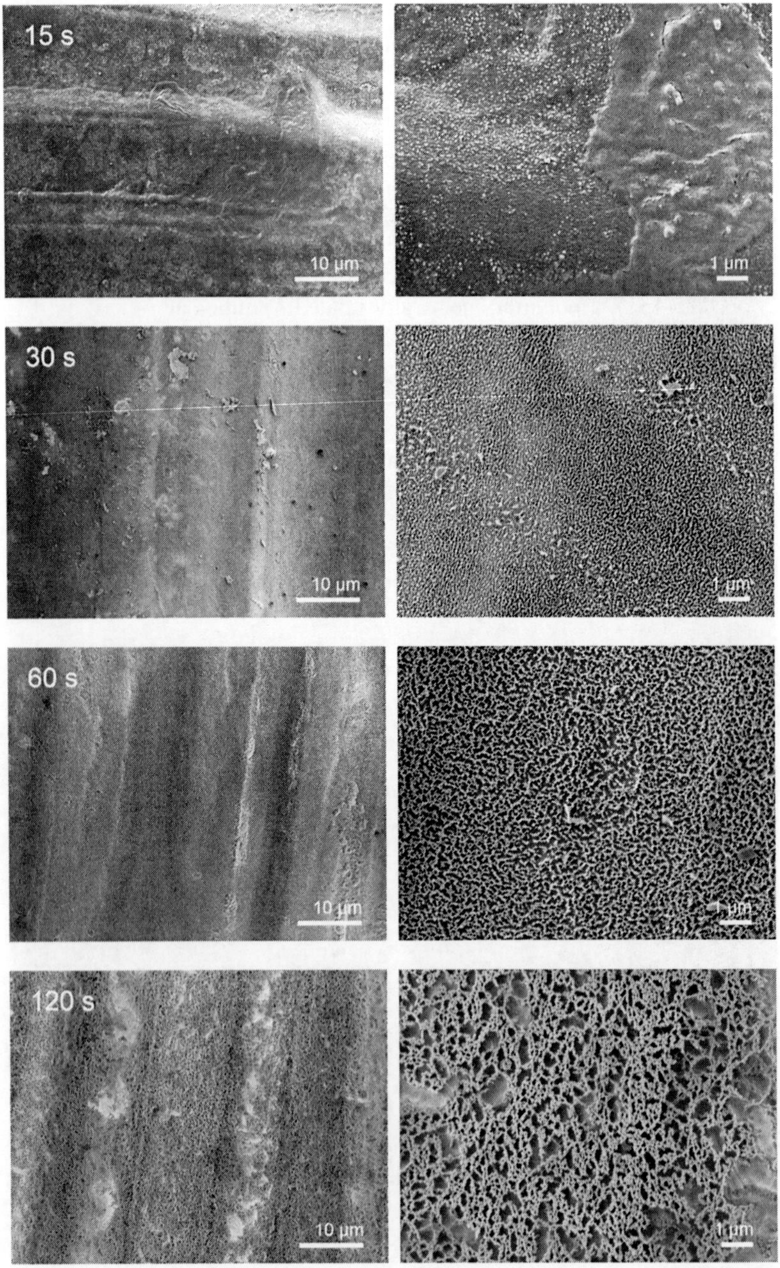
15 s
10 µm
1 µm
30 s
10 µm
1 µm
60 s
10 µm
1 µm
120 s
10 µm
1 µm

Figure 9.1 (previous pages): SEM images, unpeeled clove, pilot-scale E-mode glow. P_N = 200 W, p = 50 Pa, t = 0 – 120 s. Top to bottom: untreated, 5 s, 15 s, 30 s, 60 s, 120 s. Taken at 2,000 × (left) and 10,000 × (right) magnification. Found on previous pages.

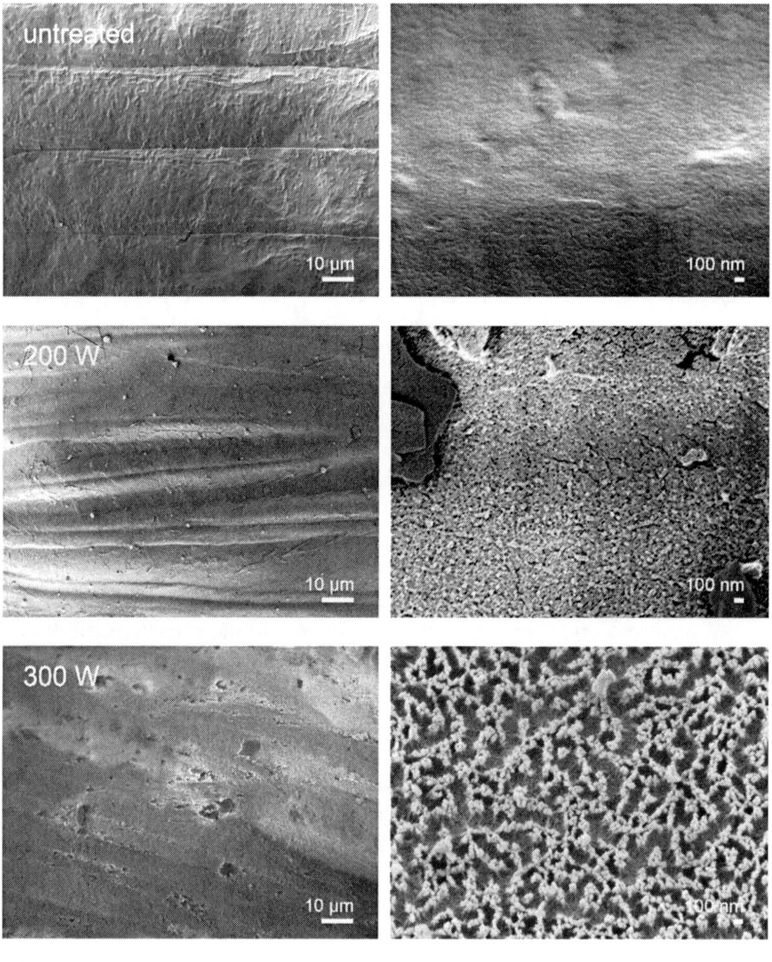

Figure 9.2: SEM images, unpeeled clove, pilot-scale E-mode glow, pulsed. Untreated; (200 W) P_N = 200 W, p = 70 Pa, t = 30 s (6 × 5 s on, 5 s off); (300 W) P_N = 300 W, p = 70 Pa, t = 60 s (6 × 10 s on, 5 s off). Taken at 1,000 × (left) and 30,000 × (right) magnification.

In the case of pulsed treatment at two different discharge conditions (Figure 9.2), as expected, the etching at the condition with higher P_N and longer treatment time (300 W, 6 × 10 s on, 5 s off) is a lot more extensive than at the less intense one (200 W, 6 × 5 s on, 5 s off). Further, in comparison with the continuous treatment (Figure 9.1), the pulsed treatment at 200 W shows slightly less intense etching at the same total treatment time (30 s). However, this is expected, as the pulsed treatment here is performed at higher working pressure (70 as opposed to 50 Pa), where the O atom density at P_N = 200 W is somewhat lower (around 6×10^{21} m^{-3}, as opposed to 7×10^{21} m^{-3}).

9.3.2 Pilot-scale afterglow plasma

In contrast to glow plasma treatment, the etching at the afterglow position is practically exclusively chemical in nature. Even though the shift of plasma from E- to H-mode causes a dramatic increase in charged particle density (Zaplotnik et al. 2011, 55001), the ions are immediately lost on the reactor surfaces (Mozetič et al. 2015, 015026), and do not reach the remote sample. In addition to the oxygen atoms flowing downstream from the H-mode plasma to the sample position, some atoms are also formed at the sample position itself due to the stray capacitive coupling still present throughout the tube. The combined effect of these atoms is the chemical etching seen in the afterglow position.

SEM images for unpeeled clove samples treated in the pilot-scale H-mode plasma afterglow are shown. To compare the effects of discharge parameters on the extent of the etching, let us consider two discharge conditions: (1) P_N = 200 W, p = 50 Pa (Figure 9.3) and (2) P_N = 300 W, p = 30 Pa (Figure 9.4). As seen in Figure 6.8, the oxygen atom density at the sample position is around 2.5 – 3 × 10^{21} m^{-3} for the 300 W, 30 Pa condition. While the oxygen atom density for the 200 W condition is not known, we know that due to lower P_N and higher pressure, the value is lower than at the 300 W condition. This is reflected in the morphological changes seen in the SEM images. After afterglow treatment at the 200 W condition for 30 s (Figure 9.3), the morphology is virtually unchanged from the untreated sample seen in Figure 9.1. After 120 s, etching is apparent, but still not to the degree seen in the glow plasma treated samples at much shorter treatment times. On the other hand, the afterglow treatment at the 300 W condition causes clearly visible etching already after 15 s of treatment (Figure 9.4). The etching appears to progress with treatment time, but is uneven across the clove surface area.

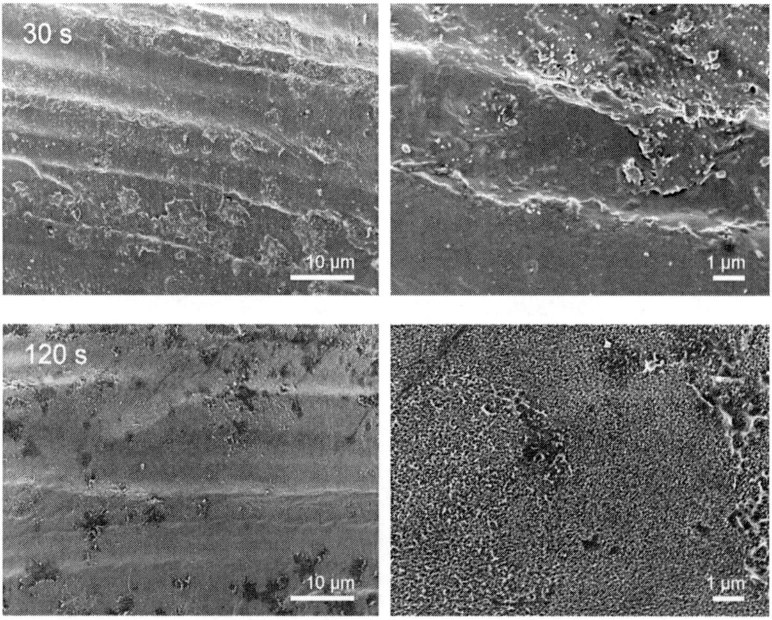

Figure 9.3: SEM images, unpeeled garlic, pilot-scale H-mode afterglow. $P_N = 200$ W, $p = 50$ Pa, $t = 30$ s, 120 s. Taken at 2,000 × (left) and 10,000 × (right) magnification.

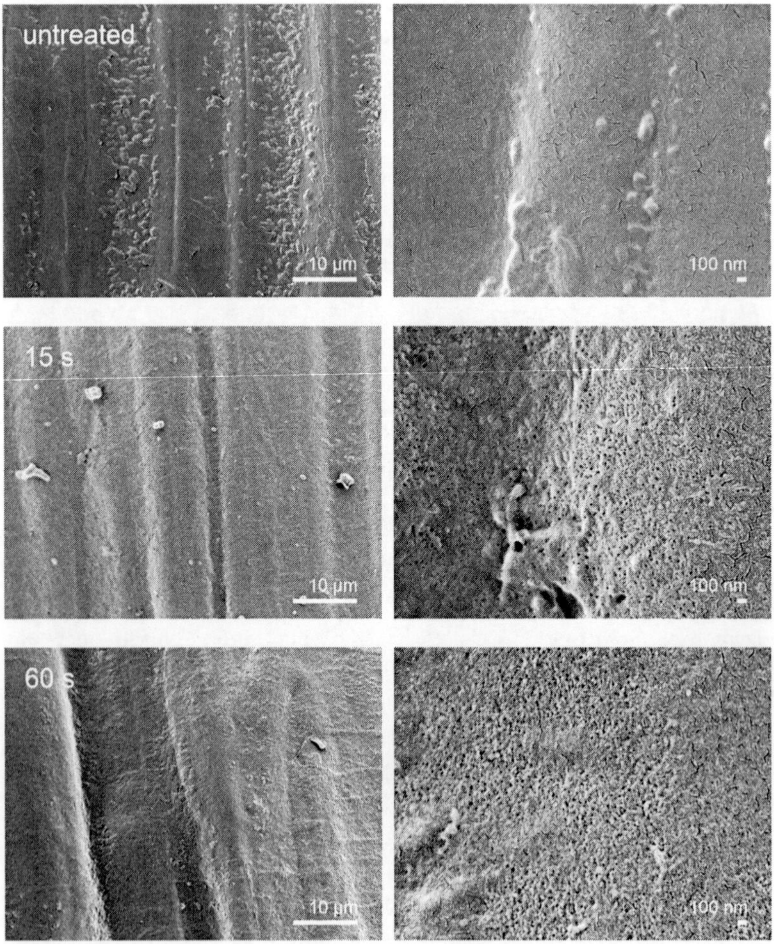

Figure 9.4: SEM images, unpeeled garlic, pilot-scale H-mode afterglow. $P_N = 300$ W, $p = 30$ Pa, $t = 0 - 60$ s. Taken at 2,000 × (left) and 30,000 × (right) magnification.

9.4 Clove surface

9.4.1 Pilot-scale glow plasma

As would be expected, there is a difference between the unpeeled and peeled clove surface also on the morphological level. The untreated peeled clove surface, shown in Figure 9.5, appears wavy at low magnification. Elongated, roughly rectangular shapes with raised ridges and concave surfaces cover the area, arranged in irregular rows. These are the outlines of the cells of the outer epidermis of the storage leaf, as evident from their size (approximately 70 × 25 μm) and shape (Mann 1952, 195-251). Similar as the protective leaf surface, the cells give shape to the cuticular layer covering them.

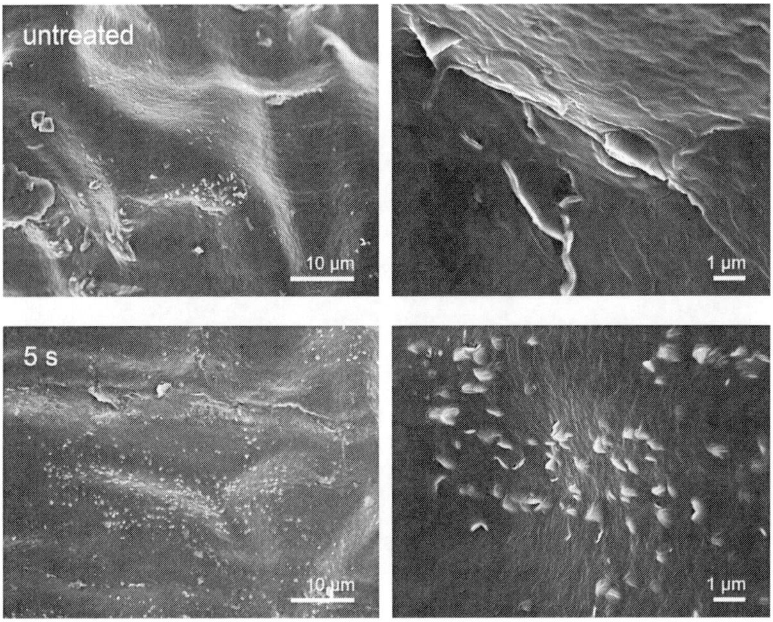

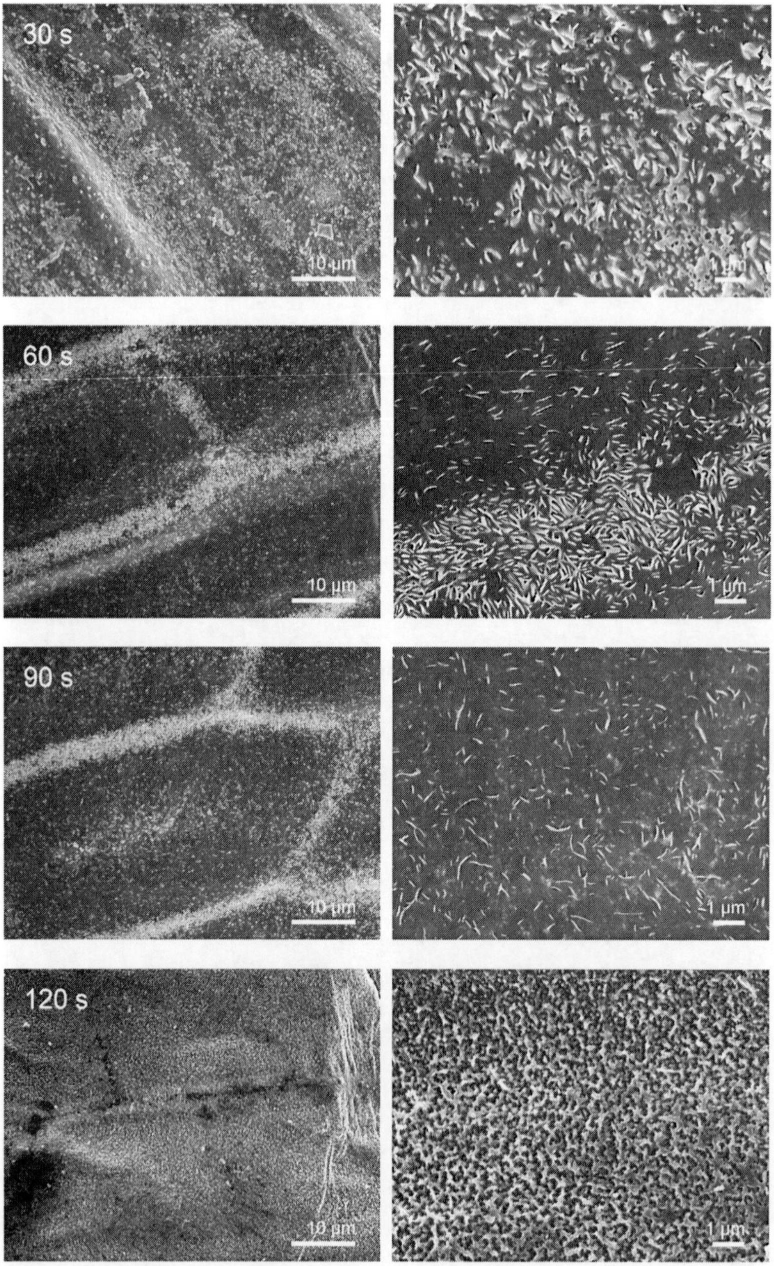
30 s
10 µm
1 µm
60 s
10 µm
1 µm
90 s
10 µm
1 µm
120 s
10 µm
1 µm

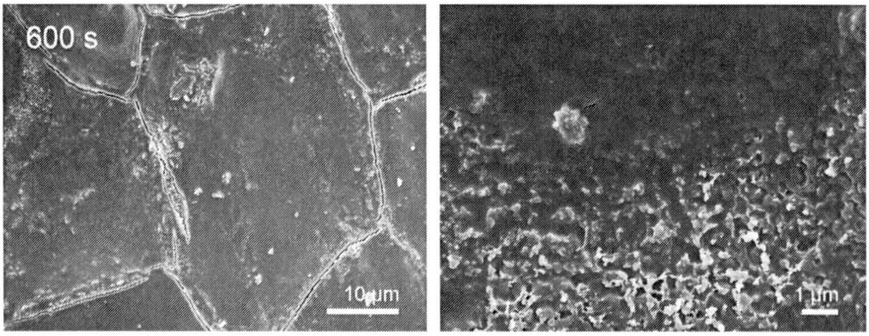

Figure 9.5: SEM images, peeled garlic, pilot-scale E-mode glow.P_N = 200 W, p = 50 Pa, t = 0 – 600 s. Top to bottom: untreated, 5 s, 30 s, 60 s, 90 s, 120 s, 600 s. Taken at 2,000 × (left) and 10,000 × (right) magnification.

At higher magnification, the surface appears wrinkled, but reasonably smooth, similar to the surface of the protective leaf. After as little as 5 s of treatment in the pilot-scale E-mode glow plasma, distinct structures begin to appear at the clove surface. They are reminiscent of flat semicircular disks protruding from the cuticle surface, and fully achieve this shape during the first 60 s of plasma treatment. Approximate size of the structures is up to 500 nm in length and up to 100 nm in width. We may visually estimate that their height is no larger than their length, up to a few hundred nm. Their number and density changes with increasing plasma treatment time. When their density increases, most prominently at 60 s of treatment and longer, they begin to blend and create a mesh-like structure. Their density first increases at the physically exposed parts of the surface – that is, on the raised ridges dividing the individual cells. At 90 s of treatment, they more densely cover the areas between the ridges, and by 120 s, they have covered virtually the entire clove surface, creating a continuous, rough nanostructure. By 600 s, the surface appears devoid of the structures, which seem to have flattened or crumpled, blending in with the remaining area.

Considering this evolution of the wax morphology with treatment time, it appears as if the structures are not formed by etching away the material around them, but rather protrude from the existing surface. The clove surface area not covered by the wax structures does not become rougher or more wrinkled with increased treatment time. This is in stark contrast with the extensive etching seen at the surface of the protective leaf (Figure 9.1). In both cases, the untreated cuticle appears as a flat film, yet the outcome of plasma treatment at the same discharge conditions is dramatically different. After longer treatment (120 s, 600 s), it appears as if etching begins to take place in parallel to the reshaping. The morphological changes become uniform across the surface, and the irregular structures typical for etching appear. To explain this delay in the etching process compared to the unpeeled clove, we also need to consider the more intense water loss from the peeled clove. One explanation is that the water lost during peeled clove treatment is adsorbed water, normally prevented from evaporation by the protective leaf. If, however, the source of the water is the bulk of the clove, it is possible that water is lost through small cracks in the cuticle. As seen in Figure 9.5, the untreated peeled clove surface is completely smooth even at magnifications as large as 30000 ×. After as little as 5 s of plasma treatment, cracks appear at the surface. If these cracks increase the porosity of the cuticle and allow for water loss, this explains why the plasma characterization shows more OH radicals during peeled clove treatment. This is in agreement with findings on plasma-treated agricultural seeds, where cracks in the cuticle surface have often been detected. Examples include plasma-treated pea (Stolárik et al. 2015, 659-676), *Andrographis paniculata* (Tong et al. 2014, 260), cotton (Wang et al. 2017, 5601), or wheat seeds (Šerá et al. 2010, 2963-2968), where cracks in the surface are interpreted as signs of improved water permeability of the seed coat.

9.4.2. Industrial-scale plasma

As explained, the treatment in the industrial-scale plasma reactor is gentler as a consequence of lower power density in the larger reactor volume. This results in a lower density and different distribution of the observed wax structures after industrial-scale plasma treatment of peeled cloves (Figure 9.6). The structures are similar, yet somewhat larger (up to 1 μm in length), than the ones after treatment in the pilot-scale plasma. They are visually most similar to what is seen in the pilot-scale plasma after 5 – 30 s of treatment (Figure 9.5). There are also differences in their density and distribution. They do not appear primarily on ridges, but are seemingly distributed randomly, and appear in discreet areas reminiscent of islands or stripes. In areas where they are present, their density appears uniform. While it is difficult to estimate the total covered area, it appears to increase with decreasing pressure, suggesting that the structures are more abundant after treatment with a higher reactive species dose, in parallel with increasing treatment time in the pilot-scale plasma.

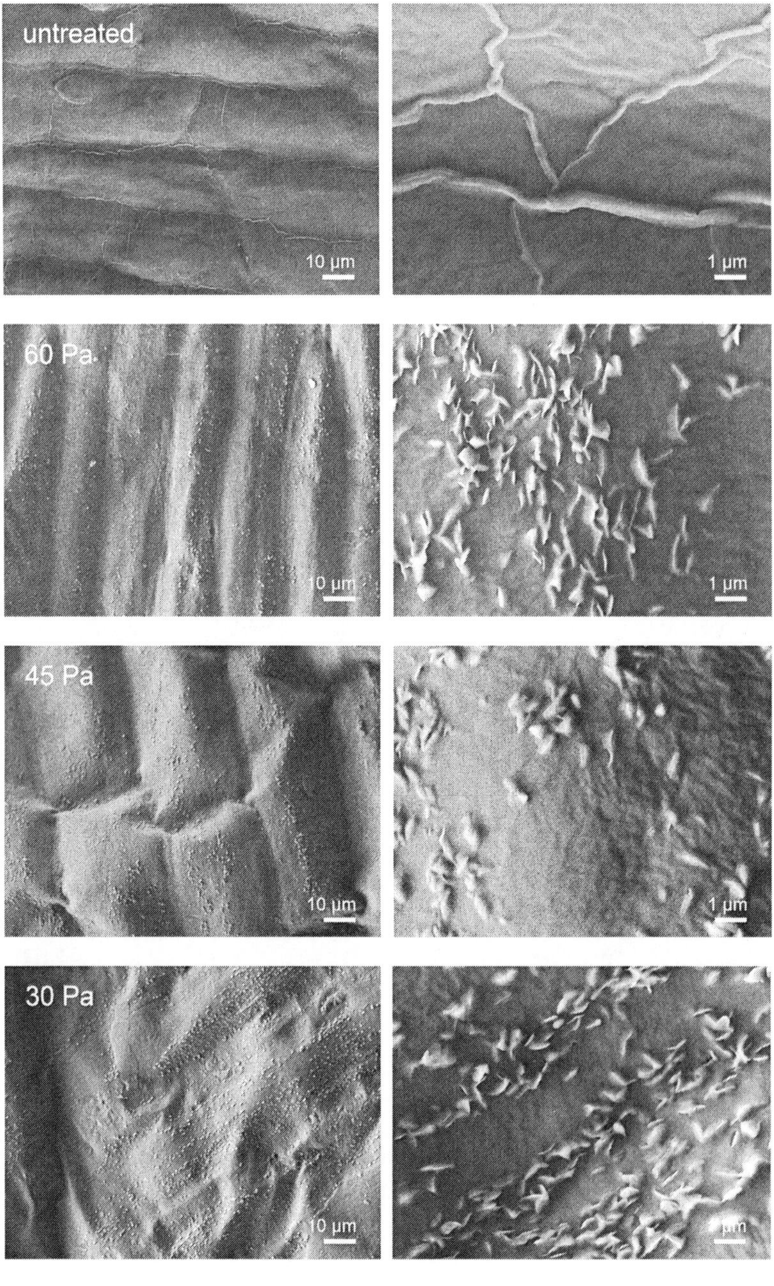
untreated
10 µm
1 µm
60 Pa
10 µm
1 µm
45 Pa
10 µm
1 µm
30 Pa
10 µm

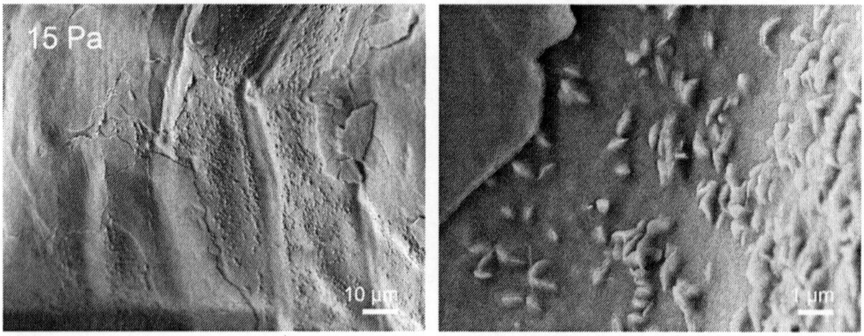

Figure 9.6: SEM images, peeled garlic, industrial-scale E-mode glow. $P_N = \sim 700 - 900$ W, $p = 15 - 60$ Pa, $t = 60$ s. Top to bottom: untreated, 60 Pa, 45 Pa, 30 Pa, 15 Pa. Taken at 1,000 × (left) and 10,000 × (right) magnification.

9.5 Surface topography

Clove surface topography is shown in AFM images of untreated and plasma-treated clove surfaces. AFM scans of protective leaf surface (Figure 9.7) demonstrate increasing surface roughness with increasing treatment time and PN in the pilot-scale E-mode glow pulsed regimen. The surface shows subtle roughness increases after the 200 W treatment, but intense nanostructuring after the 300 W treatment, just as in the corresponding SEM images (Figure 9.2).

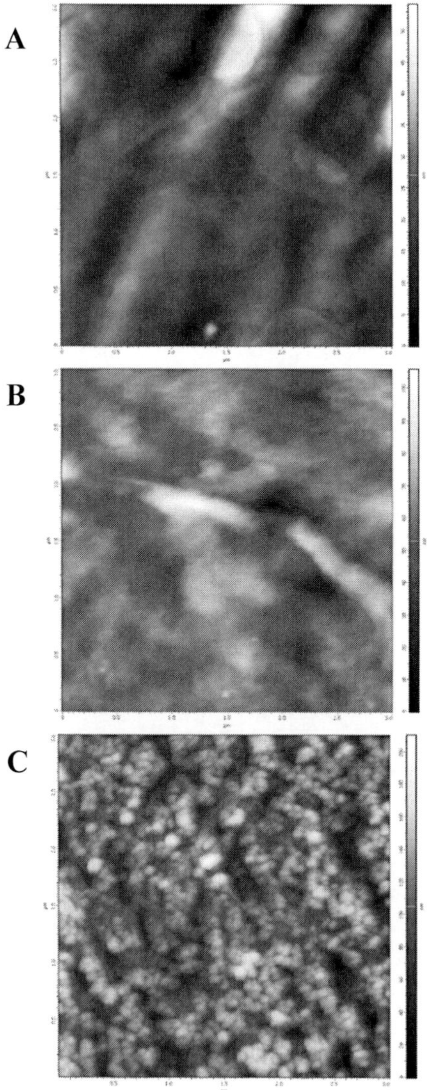

Figure 9.7: AFM images, unpeeled clove, pilot-scale E-mode glow, pulsed. (A) untreated; (B) P_N = 200 W, p = 70 Pa, t = 30 s (6 × 5 s on, 5 s off); (C) P_N = 300 W, p = 70 Pa, t = 60 s (6 × 10 s on, 5 s off). Analysis area: 3 × 3 μm.

As seen in AFM images of peeled clove surfaces (Figure 9.8), roughness of the surface noticeably increases industrial-scale plasma treatment. Waxy structures, as seen in SEM images (Figure 9.5, Figure 4.6), are also visible at the sample surface (Figure 9.8 (E)). As such, AFM scans of both the protective leaf and peeled clove surface confirm the SEM findings for the respective plasma treatment used.

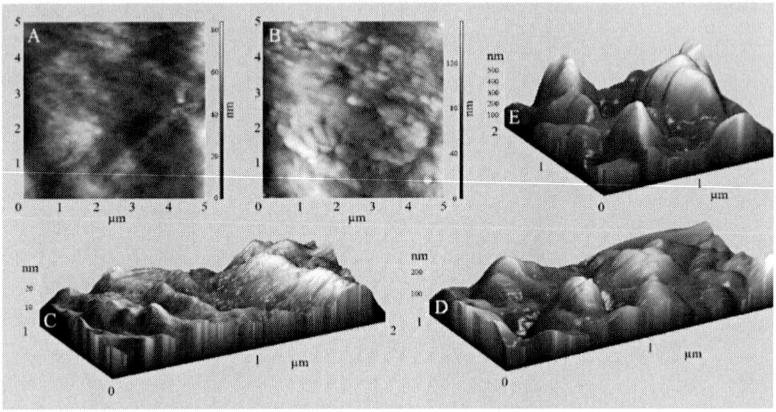

Figure 9.8: AFM images, peeled garlic, industrial-scale treatment. (A, C) untreated; (B, D, E) P_N = 800 W, p = 30 Pa, t = 60 s. Analysis area: (A, B) 5 × 5 μm, (C, D) 2 × 1 μm, (E) 2 × 2 μm.

9.6 Surface reshaping processes

9.6.1 Chemical etching

Chemical etching occurs in particular after plasma treatment of unpeeled cloves. Pure oxygen plasmas are especially efficient in the removal of organic material from surfaces (Korzec et al. 1994, 369-378), since oxygen atoms are particularly useful for oxidative degradation of organic materials due to their number of valence electrons and high electronegativity (Egitto et al. 1990, 321-422). As discussed earlier, oxygen atoms are the dominant reactive species in weakly ionized oxygen plasmas like the ones employed here. This suggests that chemical etching, and not ion bombardment, is the prevalent mechanism of material removal from the surface. In this mechanism, the wax components are oxidatively degraded by oxygen atoms, which first oxidize the structure by the functionalization mechanisms presented earlier, and then oxidize it further to form CO_2 (via formaldehyde and carbon monoxide (Dufour et al. 2014, 224007)) and H_2O. These molecules are then removed from the surface due to their volatility. This was previously confirmed for low pressure oxygen plasma-treated lamb's lettuce leaves using FTIR (Grzegorzewski et al. 2010, 1145-1152) as well as for tomato fruit cuticles using nuclear magnetic resonance (Zhang et al. 2013, 425-432). This also correlates with surface chemistry analysis results showing gradual oxidation of the cuticle surface (Table 8.2).

9.6.2. Etching depth

Due to the surface-specific nature of its effects, we expect the plasma treatment to only affect outer cuticle layers, and to not have any direct physico-chemical impact on the cells of the underlying epidermis. This is also suggested by the ATR-FTIR results, as explained above. To gain insight into which layers are etched away by the plasma treatment and which are exposed, we need to consider the etching depth. It is not possible to measure the etch rate of the garlic clove surface directly by clove mass measurements due to simultaneous loss of water from the clove, which greatly impacts the total loss of clove mass during plasma treatment. However, we can estimate the etching depth from available data for related compounds. Polyethylene foil etch rates in a capacitively coupled RF oxygen plasma reach up to about 8 μg cm^{-2} min^{-1} (Sprang et al. 1995, 689-695). If, based on the densities of polyethylene and paraffin wax, we assume a cuticular wax density of $\rho = 1$ g cm^{-3}, we can calculate an etching depth of up to 80 nm per minute:

$$h = \frac{V}{s} = \frac{m}{\rho s} = \frac{kt}{\rho} = \frac{8 \times 10^{-6}\,\mathrm{g\,cm^{-2}\,min^{-1}} \times 1\,\mathrm{min}}{1\,\mathrm{g\,cm^{-3}}} = 80\,\mathrm{nm}. \tag{9.1}$$

Here, h is the etching depth, s is the surface area, and k is the etch rate. As plant cuticles can be as thin as 100 nm (Riederer and Müller 2006), it is entirely plausible that any treatment longer than a minute could etch away the cuticle completely and expose the underlying layers of pectin and the epidermal cell walls. However, this scenario assumes a high etch rate as well as the thinnest possible cuticle. Linear hydrocarbon polymers are in general not as easily chemically etched due to the absence of tertiary C atoms, which are preferential targets for hydrogen removal by O atoms (Patiño et al. 1995, 159-171), and thereby for initiation of the etching mechanisms presented above. Considering this, along with ATR-FTIR results, it appears much more likely that the cuticle is not completely etched away by the plasma treatment, but remains at least partly intact, and that the structures observed in SEM images are either cuticular components or directly connected to the cuticle. Such a scenario was also suggested for low-pressure RF oxygen plasma treatment of tomato cuticles, where the epidermis was deemed to remain intact in the experiment, with the possibility to be damaged with prolonged exposure times (Zhang et al. 2013, 425-432).

Regardless of the chemical nature of the material, what we observe may be the selective plasma etching of cuticular components, which is a well-known concept for polymers (Puliyalil and Cvelbar 2016, 108). In oxygen plasma, polymer etch rate depends on the energy required to break the chemical bonds of the polymer backbone. Some polar and aromatic functional groups appear to stabilize the polymer against plasma etching (Taylor and Wolf 1980, 1087-1092), while on the other hand, ester bonds (such as in the cutin structure) seem to increase the degradation rate of hydrocarbons in oxygen plasma, possibly due to resonance stabilization of radicals formed in this way (RCO•, RCOO•) (Clouet and Shi 1992, 1955-1966). We assume what we see in the SEM images is a distinct cuticle component found in the deeper cuticle layers which is less readily etched, or remains in place and is slowly oxidized further. This is impossible to confirm by the performed XPS analyses, since all cuticular components, as well as pectin and cell wall components, including lignin, are strictly oxygenated hydrocarbons. ATR-FTIR also does not give useful information in this regard due to its analysis depth, which extends far beyond the layers of interest in this case. However, this idea could be correlated with the rapid increase of the surface O concentration in the first 30 s of plasma treatment and the much more gradual increase thereafter. Based on the scaffold-like structure seen after etching, the less-etched component could be cutin. If the etching depth is higher, another possibility is lignin, a polyphenolic polymer which deposits in the plant cell wall, between the cellulose micro-fibrils, where as an additional protective measure, it strengthens and waterproofs the cell wall (Barros et al. 2015, 1053-74). In the protective leaf outer epidermis, lignification does occur, and the cells thereby form a continuous hardened layer, which is then covered by the cuticle (Mann 1952, 195-251). Due to its chemical structure, specifically its many aromatic rings, we expect lignin to be less susceptible to degradation by oxygen plasma than cuticular waxes (Taylor and Wolf 1980, 1087-1092).

9.6.3. Wax structure formation

To elaborate on the origin of the wax structures that appear on the peeled clove surface after plasma treatment, we can look into cuticular wax chemistry and morphology. On many plant cuticle surfaces, epicuticular waxes are known to readily self-assemble into crystalline structures. The morphology of these structures varies widely; it is associated with plant species and determined by chemical constituents of the wax. If such wax is deposited on an artificial surface after extraction or evaporation, it will reassemble into structures typical for its composition (Riederer and Müller 2006; Niemietz et al. 2009, 221-227). While in our findings, the cuticle of untreated protective leaf exhibits a smooth wax film, it is possible that the appearance of wax structures is brought upon by the combined action of wax oxidation and surface heating during plasma treatment. The gradual increase in the C=O and O–C=O bonds seen in high-energy-resolution XPS spectra (Figure 8.11) indicates the appearance of aldehydes/ketones and carboxylic acids. C=O groups, along with –OH groups, typically imply the presence of platelet-like structures on epicuticular wax. In fact, the waxes on leaves of onion and leek – both, like garlic, of the genus *Allium* – were shown to form wax platelets, albeit more densely packed, and having serrated edges (Riederer and Müller 2006), as opposed to the smooth circular edges seen in our case. Size of the structures appears roughly comparable. In leek, the symmetric C_{31} ketone palmitone was found a major component of epicuticular wax, and may be the substance prevalent in the wax crystals on leek leaf surface (Rhee et al. 1998, 901-911).

In Figure 7.2, we have shown that due to processes such as surface neutralization and recombination, the temperature of the clove during plasma treatment rises with plasma treatment time. The reactive plasma species exert their exothermic effects at the surface, so the temperature increase of the treated material is the highest at its surface as well. For an array of studied plant species, it was found that the cuticular wax melting point is typically around 60 – 80 °C (Ensikat et al. 2006, 45-59). However, even annealing at temperatures below the wax melting point may be sufficient for wax structure formation, as long as they allow for mobility of the wax molecules (Niemietz et al. 2009, 221-227). After treatment, the chemically altered waxes may recrystallize in the form of platelets like the ones seen in our SEM images. For comparison, at the surface of Thuringian mallow seeds, plasma treatment led to a restructuring, or “sharpening”, of the cuticle, which also improved germination parameters. However, longer exposure times also led to surface erosion and cracking, followed by weaker germination improvement (Pawłat et al. 2018, 1700064; Pawłat et al. 2018, e0194349).

Based on all of the above, we propose that the appearance of wax structures on the garlic clove cuticle, as opposed to etching, is a consequence of (1) a more delicate cuticle and epidermis anatomy, (2) altered plasma chemistry due to water loss from the clove, and (3) the resulting recrystallization of epicuticular wax. Points (1) and (2) are further elaborated below.

We should allow for the possibility that some of the distortions and cracks seen at the clove surface at high magnifications are artifacts of the sample preparation process, particularly the freeze drying step. However, we can state that the sample preparation does not result in the formation of the wax structures, as they are completely absent from the surface of the untreated control sample, prepared for SEM observations in parallel with the treated samples.

9.6.4 Differences between effects on unpeeled and peeled clove

There are several potential reasons for the different behavior of the protective leaf and the storage leaf when exposed to oxygen plasma. They can roughly be explained in terms of differences between the surface anatomy and chemistry, as well as in their influence on plasma chemistry.

First, the anatomy of the epidermis and cuticle differs slightly between the two organs. In contrast to the lignification seen in the outer epidermis of the protective leaf, the outer epidermis of the storage leaf is thin and delicate (Mann 1952, 195-251). Second, as seen from XPS analysis of the untreated surface (Table 8.1), the composition of the cuticle differs. The clove surface has a higher C concentration at the surface and a lower O/C ratio, indicating a more "waxy" composition – the presence of more non-functionalized hydrocarbons, or perhaps longer aliphatic chains in the functionalized compounds. This influences the functionalization process. At the protective leaf surface, where O concentration is higher, and cuticle waxes already containing terminal –OH groups, such waxes may undergo more rapid further oxidation, as there is no need to introduce the initial O-containing functional group onto the alkyl chain. Therefore, the chemical etching that follows functionalization also proceeds more rapidly.

Third, while the protective leaf appears almost completely dry, and all its cells except the outer epidermis collapse when the leaf is mature (Mann 1952, 195-251), there is a significant amount of water – around 65 % of the clove mass – present in the storage leaf. When the peeled clove is treated in the plasma, water is lost from the clove, as indicated by loss of clove mass during treatment (Figure 10.6) as well as the prominent presence of the OH radical band at 309 nm in the OES spectra (Figure 5.2 – Figure 5.4). In

Figure 5.2, we see that the OH band is relatively more prominent during peeled clove treatment, compared to the protective leaf treatment. Naturally, the increased presence of water influences the plasma chemistry, and further, has an effect on the etching and functionalization mechanisms. As we have noted earlier, the oxidizing effect of oxygen plasma containing some water vapor should be similar to that of pure oxygen plasma, only milder. If this contributes to the slower oxidation process of the peeled clove surface, it also results in a delayed chemical etching compared to the protective leaf surface.

10. WETTABILITY AND WATER UPTAKE

10.1 Wettability, water uptake, and germination

Water uptake is of utmost importance for plant germination. In fact, water uptake is by definition the first step of germination, and is essential to the re-start of the metabolic processes associated with it (Taiz et al. 2014). Water availability to the seed or clove is therefore essential for germination to begin.

We further know that interaction of oxygen plasma with organic surfaces typically increases the wettability of such surfaces. For that reason, it is essential to evaluate the interaction between water and seed surface, or in the case of garlic, clove surface. If this interaction can be modified by the effects of plasma, it should be possible to influence the germination process as well. In this context, it is possible to assess two characteristics: wettability (or, in the case of water, hydrophilicity) of the given surface, as well as the water uptake by the seed or clove.

10.2 Water contact angle

Water contact angle (WCA) determination is a simple, widespread technique for assessing surface wettability. In the sessile drop method, a droplet of purified water, typically on the μL order, is added onto a smooth surface. A camera with suitable software then records the angle between the surface and the tangent at the droplet edge. The lower the WCA, the more hydrophilic the analyzed surface. As wettability increases, the WCA nears zero until complete wetting when the droplet spreads across the surface. In addition, it is common to determine contact angles for liquids other than water and to calculate surface energy from the contact angle measurements.

When we consider WCA measurement results on plasma treated garlic clove surfaces (Figure 10.1), we notice that the WCA of the untreated protective leaf as well as the untreated clove surface is comparable, around 100°. With increasing plasma treatment time, the WCA decreases, as the surface gradually becomes more hydrophilic. The differences in chemical and morphological characteristics between unpeeled and peeled cloves cause them to respond differently to plasma using the same discharge parameters.

On unpeeled garlic, the WCA for cloves treated at 100 W gradually decreases, and the values are comparable regardless of working pressure (30 or 50 Pa). This is in line with the oxygen atom density recorded with the catalytic probe at these parameters, which at 100 W does not vary in the pressure range of 30 – 70 Pa. The WCA of the unpeeled clove treated at 200 W decreases more rapidly compared to 100 W, and we achieve practically complete wetting already after a treatment time of 60 s, as the droplet spreads across the surface fully.

Using the same discharge parameters, the WCA decrease on the peeled clove surface is slower. The WCA decreases approximately equally for all conditions at 100 W and 200 W, even though the oxygen atom density at 200 W is twice as large as at 100 W. The WCA decreases more rapidly, but also approximately equally, for both conditions at 300 W. From these findings, it appears that the nominal power at the RF generator is a more important parameter in influencing the garlic WCA than the working pressure. Increasing the discharge power results in an accelerated WCA decrease, and thereby, wettability increase.

We should consider that both the functionalization and the etching play a role in determining the wettability of a surface, so both affect the WCA. By introducing new, oxygen-containing functional groups onto the surface, the chemistry of the seed surface changes, and surface energy as well as wettability increase. Concurrently, an increase in surface roughness also increases the surface wettability (Fridman 2008). The two effects complement and affect each other, and the final WCA decrease is a product of both.

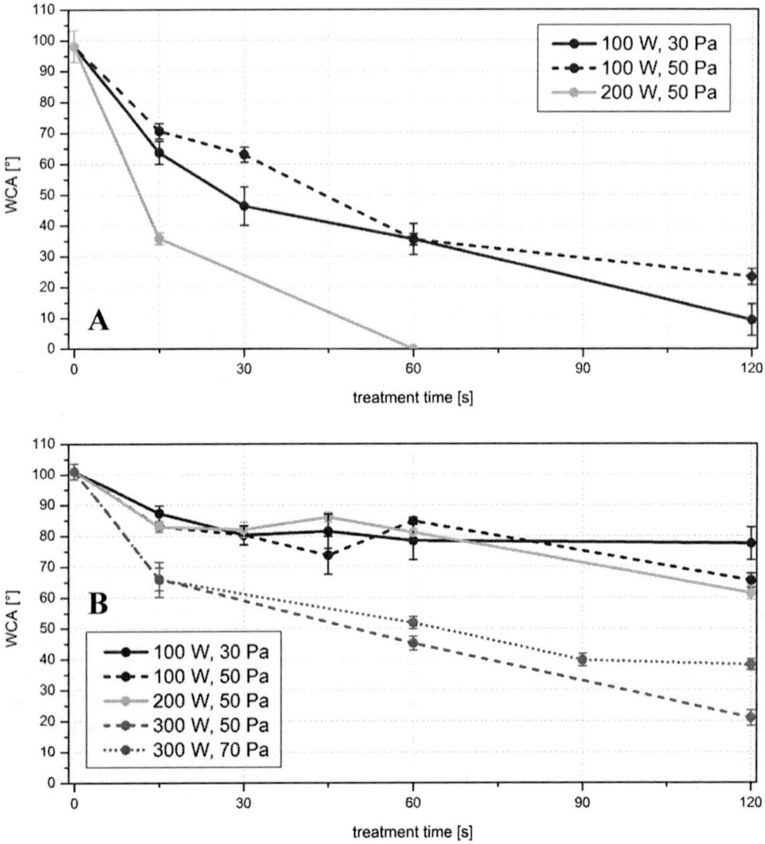

Figure 10.1: WCA on the surface of the garlic clove. (A) unpeeled (protective leaf); (B) peeled (storage leaf). Discharge parameters: pilot-scale system, E-mode glow, P_N = 100 – 300 W, p = 30 – 70 Pa, t = 0 – 120 s. The error bars represent standard error.

10.3 Water uptake

Wettability provides information regarding the interaction between water and the clove surface. However, it is important to also determine how water is taken up by the clove. To determine water uptake, cloves can be weighed before and after submerging them in purified water for a set soaking time. Water uptake is then calculated according to the equation

$$\Delta m = \frac{(m_t - m_0)}{m_0} \times 100\%, \tag{10.1}$$

where Δm is the increase in mass, expressed as a percentage of m_0, while m_0 and m_t are the clove mass immediately before and after t minutes of soaking, respectively.

10.3.1 Water uptake of untreated cloves

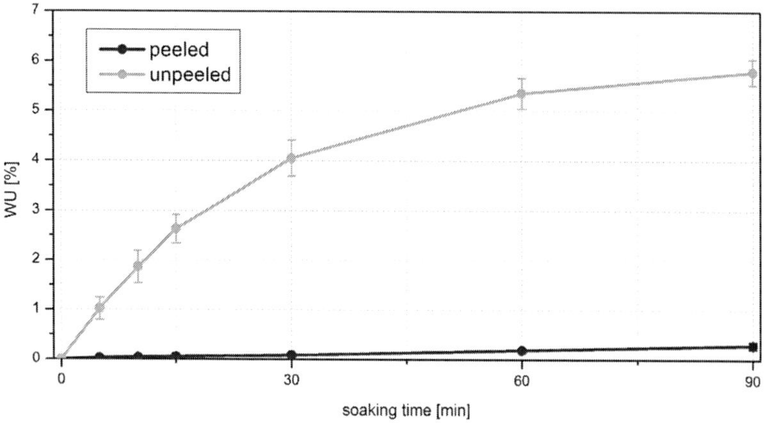

Figure 10.2: Water uptake of untreated cloves. Peeled and unpeeled, recorded after soaking for up to 90 minutes.The error bars represent standard error.

When we compare the water uptake of peeled and unpeeled garlic cloves that have not been plasma-treated (Figure 10.2), we see that unpeeled cloves take up water significantly faster than peeled ones. Contact angle measurements show that the wettability of both untreated surfaces is comparable. However, we know from the results of XPS analysis that the oxygen concentration of the protective leaf surface is several at.% higher than that of the peeled clove, and as such, is more hydrophilic than the clove surface. In addition, the protective leaf of the unpeeled clove is dry, and is the component of the clove that easily soaks up large quantities of water. It follows that when the peeled clove is soaking, the protective leaf acts as a water reservoir, taking up the water much more easily compared to the peeled clove, which is hydrophobic and does not readily take up water if not exposed to plasma beforehand.

10.3.2 Soaking time

To test the effect of soaking time on water uptake, let us consider garlic cloves plasma-treated using constant discharge parameters and a single treatment time. Such cloves are then soaked in deionized water for increasing amounts of time. Figure 10.3 and Figure 10.4 show the water uptake of unpeeled and peeled garlic cloves, respectively. It is evident that the water uptake significantly increases after plasma treatment for both unpeeled and peeled cloves. Further, as we increase the soaking time of a clove in water, the water uptake continually increases.

The water uptake increase after plasma treatment of cloves arises from the physico-chemical modifications of the surface. The functionalization of the surface with O-containing functional groups, as well as increased surface roughness, increase the wettability of the cuticle. Concurrently, the etching effect facilitates water permeation through these layers. Consequently, the surface affinity for water increases, and water penetrates into the clove more easily. The combined effect is important; while physical removal of the cuticle, such as scraping, will enhance water uptake by the seeds, it is not as effective as plasma treatment, which additionally functionalizes the surface (Srisonphan 2018, 11297-11304).

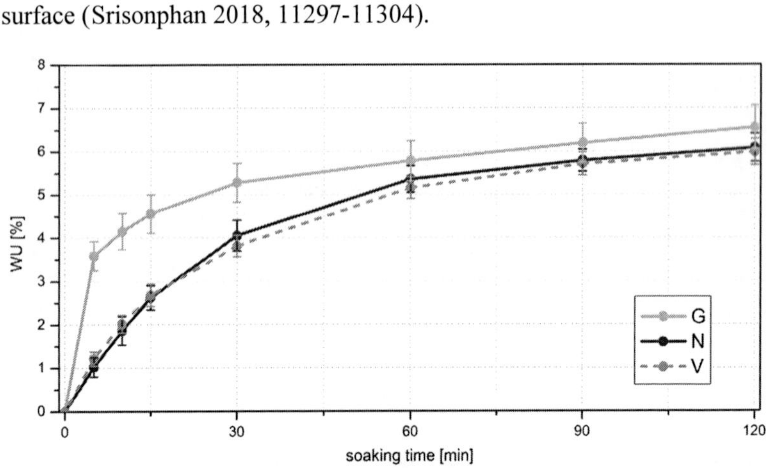

Figure 10.3: Water uptake of plasma-treated unpeeled garlic cloves. Discharge parameters: industrial-scale system, E-mode glow, $P_N = \sim 800$ W, $p = 30$ Pa, $t = 45$ s, soaking in DI water for specified amounts of time. G = glow plasma treatment; N = untreated control; V = vacuum control.

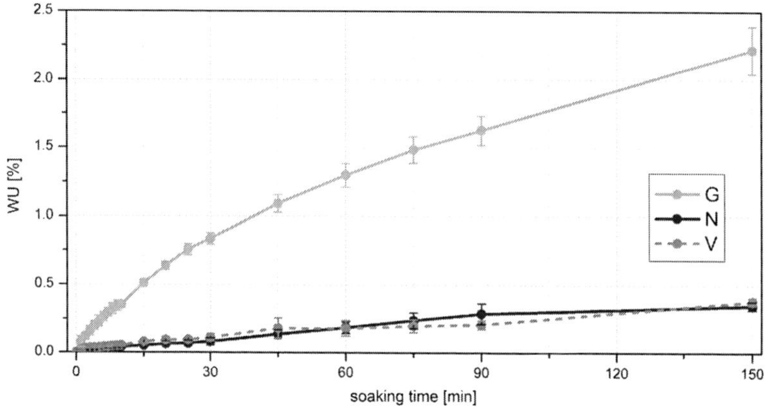

Figure 10.4: Water uptake of plasma-treated peeled garlic cloves. Discharge parameters: industrial-scale system, E-mode glow, $P_N = \sim 900$ W, $p = 40$ Pa, $t = 60$ s, soaking in DI water for specified amounts of time. G = glow plasma treatment; N = untreated control; V = vacuum control.

Just as untreated cloves, plasma-treated unpeeled and peeled cloves behave very differently regarding the specific dynamics of their water uptake. In particular, after plasma treatment, the water uptake increase for peeled cloves is much more dramatic compared to unpeeled cloves, especially with increased soaking time. However, the total water uptake relative to clove mass remains much higher for unpeeled cloves due to the water reservoir effect of the protective leaf.

The peeled clove water uptake increases steadily with soaking time. On the other hand, the water uptake of unpeeled cloves is accelerated for shorter soaking times – in this case, around 60 minutes – after which, there is no significant difference between treated and untreated cloves, as the water uptake approaches a plateau. The protective leaf mass is very low compared to the clove itself, so its capacity for water uptake is limited. Simultaneously, as we have seen in XPS results, the clove surface under the treated protective leaf is affected by the plasma treatment to a much lesser degree, meaning that the water will not penetrate such a clove more readily than the untreated one. Therefore, the water uptake increase for plasma-treated unpeeled cloves is limited, and manifests only as an acceleration in the beginning of soaking. It is reasonable to expect that a similar effect would be seen after significantly prolonged soaking of peeled cloves.

For both peeled and unpeeled cloves, there is no difference in water uptake between the untreated control and the vacuum control. While some clove mass is lost during vacuum exposure alone, this value is insignificant compared to the mass lost during plasma treatment. Further, in the absence of plasma reactive species, no surface modification takes place which could affect the WU of the clove.

10.3.3 Treatment time

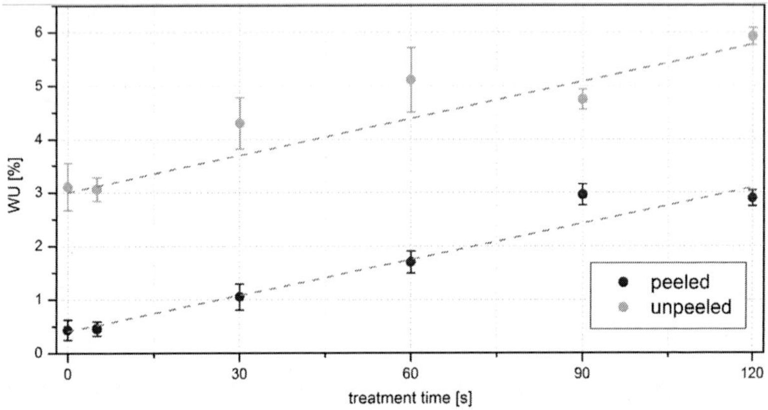

Figure 10.5: Water uptake of garlic cloves. Discharge parameters: pilot-scale system, E-mode glow, $P_N = 200$ W, $p = 50$ Pa, $t = 0 - 120$ s, soaking for 10 min. The error bars represent standard error. Linear trend lines are added.

Next, let us consider garlic cloves exposed to plasma at increasing treatment times. Figure 10.5 shows the water uptake of such cloves after treatment and soaking in deionized water for 10 minutes. With increased treatment time, the water uptake increases linearly for both peeled and unpeeled cloves. Initial water uptake of the untreated unpeeled clove is about six times higher than that of the untreated peeled clove (about 3 % as opposed to about 0.5 %). However, the slope at which the water uptake increases with treatment time is nearly identical for peeled and unpeeled cloves treated at the same discharge condition.

As with treatment in the industrial-scale system treatment, the water uptake increase corresponds to the progressive changes to the clove surface seen in the XPS, SEM and WCA analysis results, which increase the clove wettability and thereby also facilitate the water uptake.

As before, water uptake after exposure to vacuum alone is unchanged regardless of vacuum exposure within the treatment time frame.

10.3.4 Clove mass loss

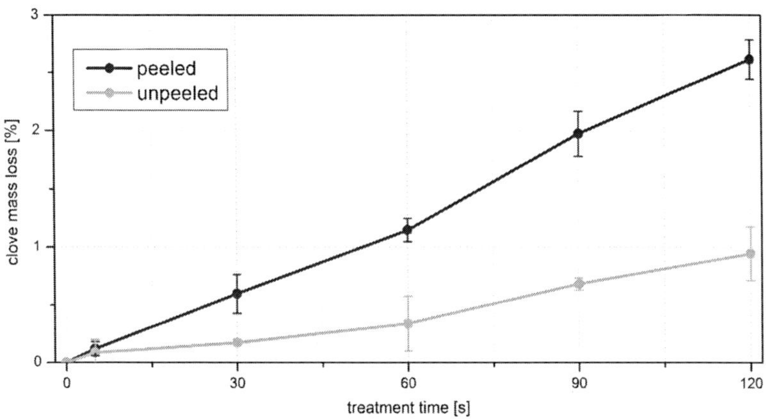

Figure 10.6: Clove mass loss for peeled and unpeeled cloves. Discharge parameters: pilot-scale system, E-mode glow, P_N = 200 W, p = 50 Pa, t = 0 – 120 s. The error bars represent standard error.

During plasma treatment, both unpeeled and peeled cloves slowly but consistently lose mass due to processes taking place at the clove surface. Loss of clove mass during treatment in the pilot-scale plasma is shown in Figure 10.6.

Some mass is already lost from cloves exposed to low pressure and gas flow alone, but that loss is minimal compared to plasma treated cloves (e.g., for peeled cloves, no more than 0.4 % at the longest treatment time). It is possible that only water adsorbed onto the clove is removed in this way. As we have explained, exposure to low pressure alone is also not sufficient for increased water uptake.

The increased mass loss during plasma treatment indicates that plasma affects the surface in a way that allows for additional loss of clove mass. The reasons for the mass loss may be either water loss from the clove bulk or surface chemical etching. During the treatment of cloves, OH radical presence in the plasma system is considerable. While OH may originate from water present in the system as well as from chemical etching, the source of CO is chemical etching alone. The difference in the intensity of peaks representing OH and CO in the OES spectra indicates that more water than carbon is lost from the clove. This suggests that during plasma treatment, the clove does indeed lose water due to the loss of protective function of the cuticle. Fresh garlic cloves contain about 65 % of water

(Kamenetsky 2007, 123-172). The plasma effects on the cuticle facilitate water penetration through this otherwise hydrophobic layer to the surface, where it gets removed from the clove due to the low pressure inside the plasma reactor. Compared to the mass of water lost in this way, material loss due to etching is likely negligible.

As seen in Figure 10.6, the clove mass steadily decreases during plasma treatment, but the loss is higher for peeled than unpeeled cloves. This is due to the function of the protective leaf, which is a barrier to the clove against water loss. Plasma treatment in the absence of any cracks or damages only affects the protective leaf surface, without affecting the surface of the clove itself. Loss of clove mass is no larger than about 2.5 % for treatment times up to 120 s in the pilot-scale plasma, which is the longest feasible treatment time for any practical application in agriculture.

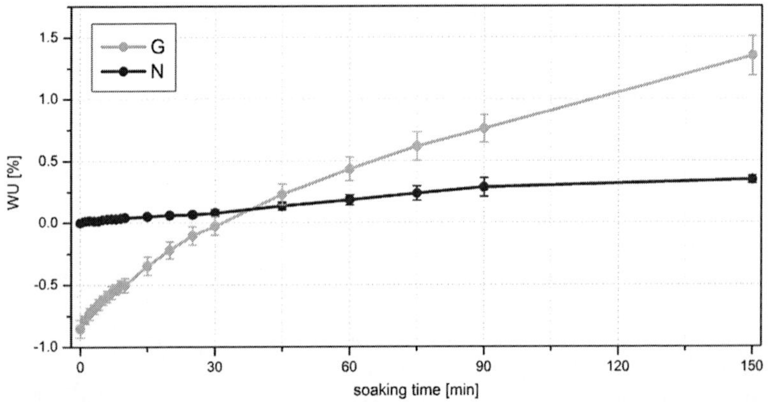

Figure 10.7: WU of peeled cloves, calculated using the initial clove mass. Industrial-scale plasma, $P_N = \sim 900$ W, $p = 40$ Pa, $t = 60$ s, soaking in DI water for specified amounts of time. G = glow plasma treatment; N = untreated control. The error bars represent standard error.

Another consequence of clove mass loss that needs to be considered is that during soaking, the water uptake first replaces the mass lost during treatment. This is illustrated in Figure 10.7, where the water uptake is calculated using the clove mass before plasma treatment in the calculation (m_0 in Eq. (10.1)). In this case, the clove mass during soaking reaches the initial clove mass only after about 30 minutes. However, even beyond that point, WU is accelerated compared to untreated cloves. This confirms that the water uptake increase or acceleration after plasma treatment is not just due to water loss from the clove during treatment, but due to changes to the surface chemistry and morphology discussed earlier.

In experiments performed on beans, it was found that the micropylar region is particularly important for seed imbibition, but nonetheless, plasma treatment strengthens water permeation through the entirety of the seed coat surface (Bormashenko et al. 2015, 4013-4021). While no micropylar region exists in the clove, a similar cavity-like structure at the tip of the storage leaf could contribute to the steady mass loss during clove exposure to vacuum. Plasma exposure then lessens the protective function of the cuticle and allows for additional water loss through the clove surface.

10.3.5 Effect of initial clove mass

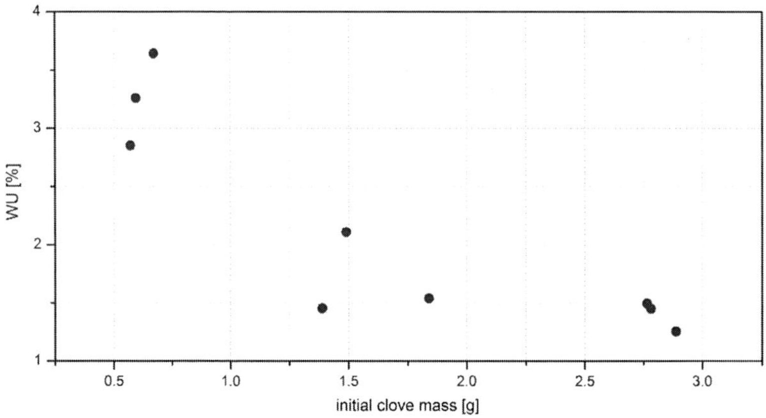

Figure 10.8: Effect of initial clove mass on WU. Pilot-scale E-mode glow, $P_N = 200$ W, $p = 50$ Pa, $t = 60$ s. The data points represent individual values.

Water uptake by the clove is also affected by the initial mass of the clove. This is due to the surface-to-mass ratio, which decreases with increasing clove mass. A heavier clove therefore has a proportionally smaller surface area through which water can be taken up into the clove. This means that at the same discharge conditions and treatment time, the water uptake of lighter cloves is proportionally larger. The relationship between the water uptake and initial clove mass is shown in Figure 10.8; all cloves are treated at identical discharge parameters and treatment time. As initial clove mass affects the water uptake, it is important to select cloves within a predefined, narrow mass range for each set of water uptake experiments.

11. SPROUT AND ROOT GROWTH

11.1 Clove growth in a laboratory setting

This chapter examines how plasma treatment can positively affect sprout and root growth of garlic cloves. A preliminary example of positive pilot-scale plasma influence on garlic sprout length is shown in Figure 11.1.

Figure 11.1: Peeled clove sprout growth, pulsed treatment. Pilot-scale E-mode glow plasma, $P_N = 200$ W, $p = 70$ Pa, $t = 30$ s (6 × 5 s on, 5 s off). N = untreated control; P = plasma treated. Recorded after 140 h of growth.

To monitor and record sprout length, root length, and root number, garlic cloves may be grown in a laboratory setting. Sprouts are measured from their lowest point of emergence from the garlic storage leaf up to the tip. Length of the longest root is recorded from the bottom of the clove to the root tip, and total number of roots is counted for each clove.

To begin germination and growth, conditions suitable for these processes need to be provided. In this case, each clove was placed on a layer of filter paper inside an open petri dish immediately after plasma treatment and moistened with deionized water. Cloves were incubated at room temperature in the dark, and water was added as needed during sampling.

11.2 Untreated controls

First, let us compare sprout and root growth of untreated controls (Figure 11.2). We see that sprouts and roots of unpeeled cloves grow significantly faster than those of peeled cloves. As already mentioned, the protective leaf is a water reservoir, and is aligned to the storage leaf tightly. Therefore, during laboratory growth of unpeeled cloves, the protective leaf takes up water, making it more directly available to the storage leaf, which then takes it up slowly over a large area. Even if this is not reflected in the water contact angle measurements (Figure 10.1), the untreated storage leaf itself is noticeably more water-repellent, and therefore the contact area between the storage leaf and water is smaller in the case of untreated peeled cloves. This further results in less water availability, followed by slower sprout and root growth.

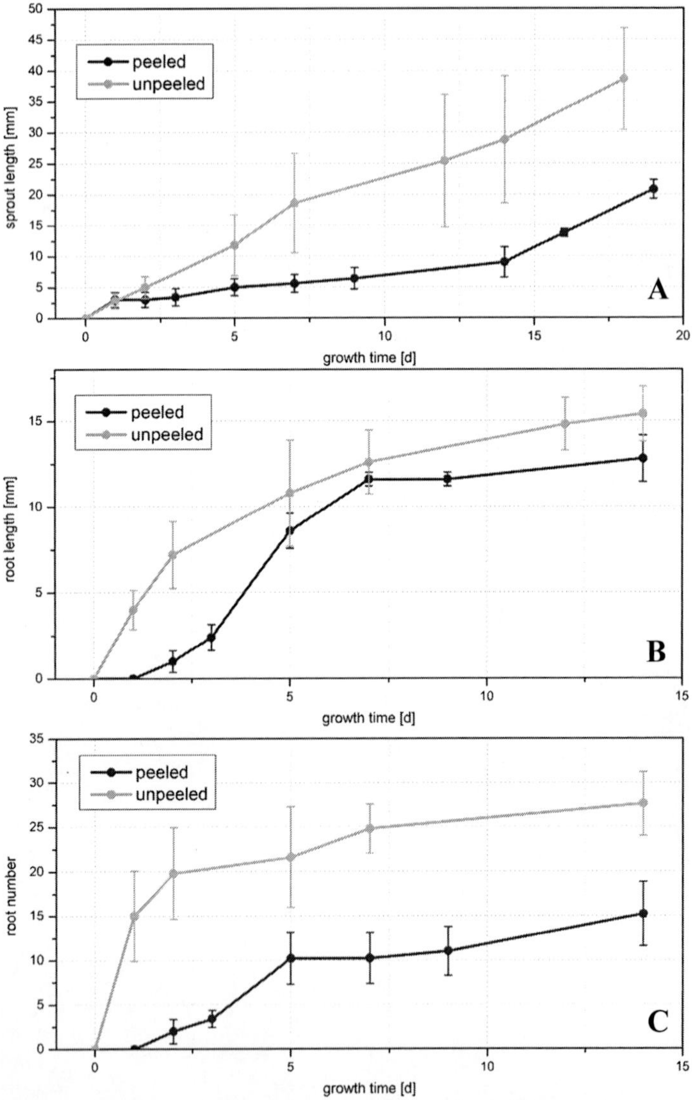

Figure 11.2: Growth of untreated clove sprouts and roots. The error bars represent standard error. (A) sprout length; (B) root length; (C) root number.

11.3 Unpeeled cloves

11.3.1 E-mode glow

To understand the effects of plasma treatment on the sprout and root growth, we need to compare the effects of various discharge parameters, as well as the plasma characteristics that result from those treatment conditions. First, let us consider the growth of unpeeled clove sprouts and roots after treatment in the pilot-scale E-mode glow plasma, specifically the effect of increasing generator power and treatment time at a constant working pressure.

Sprout growth in shown in Figure 11.3. At $p = 50$ Pa, the treatment at $P_N = 100$ W does not significantly affect sprout growth. The treatment at $P_N = 200$ W for a treatment time of 60 s significantly improves sprout growth from 12 days of growth onward. A non-significant improvement trend is also seen for $t = 45$ s. The treatment at $P_N = 300$ W does not significantly affect sprout growth, although a trend is apparent for the treatment times of 30, 45 and 60 s.

Root length is shown in Figure 11.4. We see that the treatment at $P_N = 100$ W slightly improves root length at several treatment times, but the improvement is not significant. The same is true for $P_N = 200$ W for treatment times of 15 – 60 s, as well as for $P_N = 300$ W for 15 and 30 s. In both cases, longer treatments retard root length. The negative effect is statistically significant for $t = 120$ s at both P_N.

Root number is shown in Figure 11.5. We see that root number is not significantly positively affected at any P_N or treatment time. Further, at $P_N = 200$ W and $t = 120$ s, as well as at $P_N = 300$ W and $t = 45 – 120$ s, the root number is significantly negatively affected at most growth times.

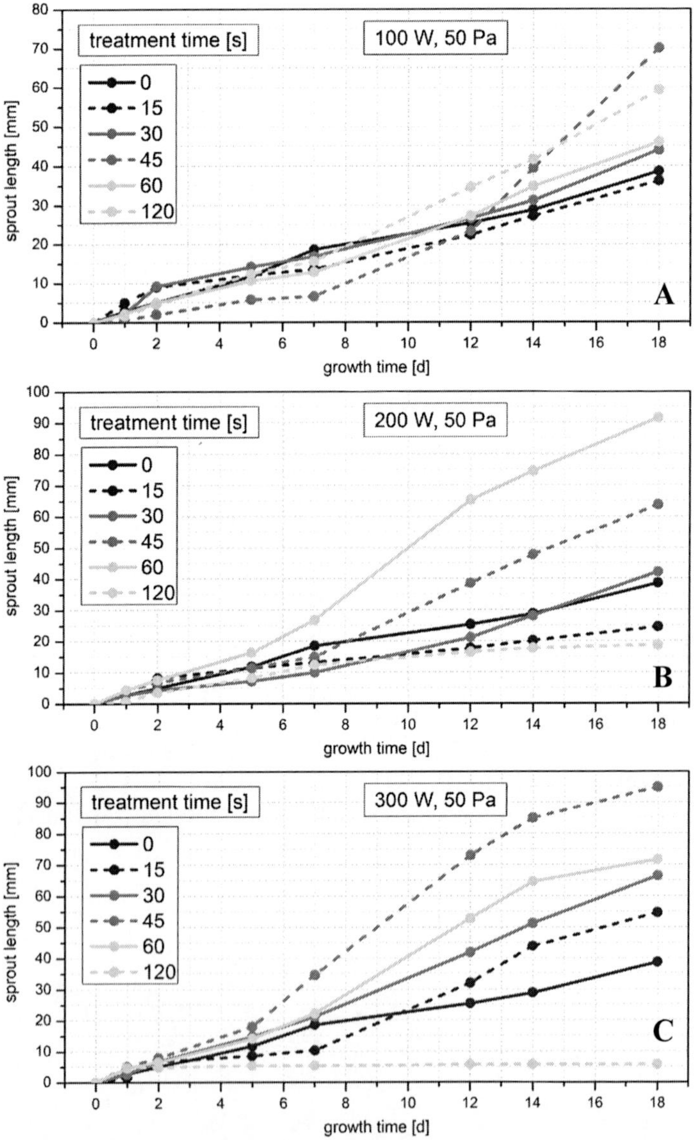

Figure 11.3: Growth of unpeeled clove sprouts, pilot-scale E-mode glow. (A) $P_N = 100$ W; (B) $P_N = 200$ W; (C) $P_N = 300$ W. $p = 50$ Pa, $t = 0 – 120$ s.

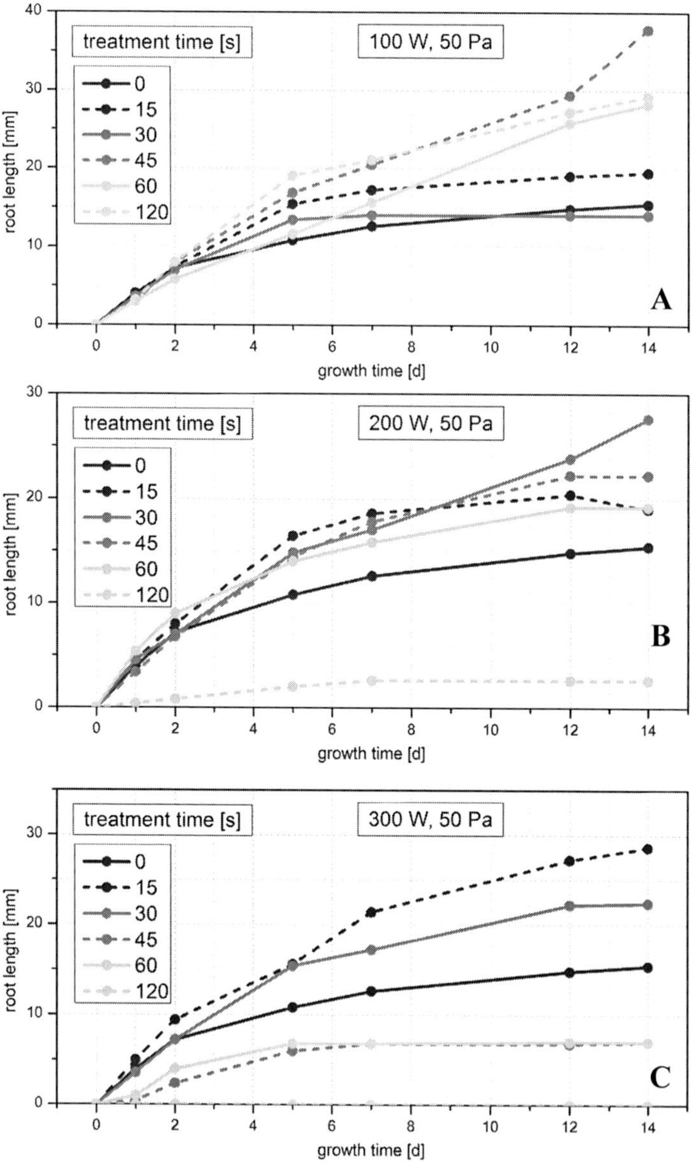

Figure 11.4: Growth of unpeeled clove roots, pilot-scale E-mode glow. (A) $P_N = 100$ W; (B) $P_N = 200$ W; (C) $P_N = 300$ W. $p = 50$ Pa, $t = 0 - 120$ s.

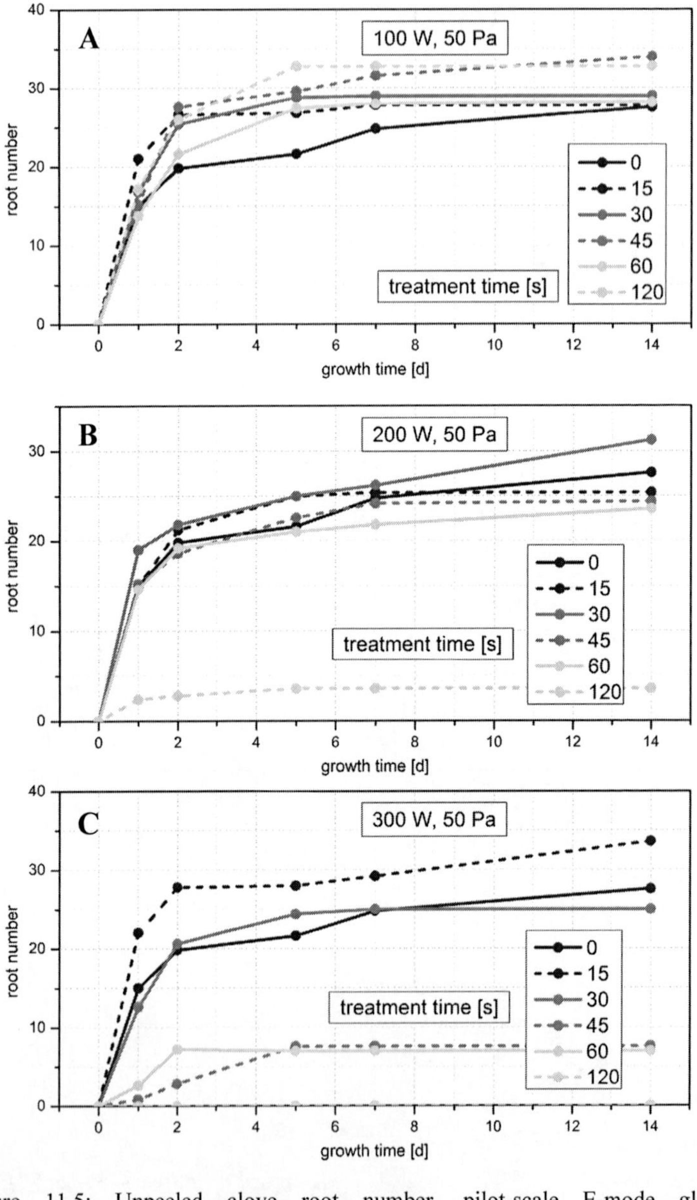

Figure 11.5: Unpeeled clove root number, pilot-scale E-mode glow. (A) $P_N = 100$ W; (B) $P_N = 200$ W; (C) $P_N = 300$ W. $p = 50$ Pa, $t = 0 – 120$ s.

From all of the above, we conclude that the treatment at $P_N = 100$ W is too weak to significantly affect either sprout length, root length, or root number. Root growth is somewhat affected, and shows a slight trend of improvement with increasing treatment time. For treatments at $P_N = 200$ W and 300 W, sprout length improvements are seen for specific treatment times. This indicates that a particular dose of oxygen atoms which the clove receives positively affects its sprout growth. We can calculate the oxygen atom dose, D, that a clove receives at the given discharge conditions from the oxygen atom flux, clove surface area, and treatment time:

$$D = j_O \, A \, t. \tag{11.1}$$

From Eq. (7.10), it follows that

$$D = \frac{1}{4} n_O < v_O > A \, t. \tag{11.2}$$

We have established in Eq. (7.4) that the surface of a clove with a mass of 1 g is about 5 cm^2, and in Eq. (7.12) that the $<v_O>$ at the given conditions is 630 m s^{-1}. The n_O at the particular discharge condition is taken from Figure 6.2, and t is the treatment time. Considering this, one gets

$$D = \frac{1}{4} n_O \times 630 \frac{\mathrm{m}}{\mathrm{s}} \times 5 \text{ cm}^2 \times t = 787.5 \, n_O t. \tag{11.3}$$

For the discharge parameters used in the present experiments, we can calculate the O atom dose received by a clove, as shown in Table 11.1.

Table 11.1: O atom dose received by a clove during treatment. Pilot-scale E-mode glow plasma, $p = 50$ Pa.

t [s]	**100 W**	**200 W**	**300 W**
0	0	0	0
15	3.5×10^{25}	8.3×10^{25}	1.1×10^{26}
30	7.1×10^{25}	1.7×10^{26}	2.2×10^{26}
45	1.1×10^{26}	2.5×10^{26}	3.4×10^{26}
60	1.4×10^{26}	3.3×10^{26}	4.5×10^{26}
90	2.1×10^{26}	5.0×10^{26}	6.7×10^{26}
120	2.8×10^{26}	6.6×10^{26}	9.0×10^{26}

If we consider which discharge conditions positively affect sprout growth (Figure 11.3), we can estimate the required O atom dose for sprout growth improvement of unpeeled cloves. The only significantly improved condition is $P_N = 200$ W, $t = 60$ s, which equals an O atom dose of approximately 3.3×10^{26}. If we include conditions that non-significantly improve sprout growth, ($P_N = 200$ W, $t = 45 - 60$ s, as well as $P_N = 300$ W, $t = 30 - 60$ s), then the required O atom dose bracket for sprout growth improvement of a garlic clove is around $2 - 4.5 \times 10^{26}$.

Following the same reasoning, we can calculate the required O atom dose for root length improvement. While none of the improvements in Figure 11.4 are significant, some of the treatments non-significantly improve root length, in particular: $P_N = 100$ W, $t = 45$ s; $P_N = 200$ W, $t = 15 - 45$ s; and $P_N = 300$ W, $t = 15 - 30$ s. This suggests that a degree of root length improvement may be achieved with an O atom dose of about $1 - 2.5 \times 10^{26}$. Conversely, from the negative effects seen at longer treatment times for $P_N = 200$ and 300 W, we can state that an O atom dose over $\sim 3.5 \times 10^{26}$ will negatively affect root length.

The required O atom dose for root number improvement is estimated less easily due to the lack of clear trends in the non-significant improvements (Figure 11.5). A rough estimate is about $1 - 2 \times 10^{26}$. As with root length, any O atom dose over $\sim 3.5 \times 10^{26}$ will negatively affect root number.

Based on all of the above, a pilot-scale plasma treatment that positively affects both sprout and root growth of unpeeled cloves in the laboratory conditions would require an O atom dose of about $2 - 2.5 \times 10^{26}$. Out of the selected treatments, this is true for $P_N = 200$ W, $t = 45$ s, as well as $P_N = 300$ W, $t = 30$ s. Of course, many other P_N/working pressure/treatment time combinations can be used to achieve this range, and are, disregarding other factors, potentially effective. It should be additionally noted that for different garlic cultivars, or even just different clove sizes, the required O atom dose for any kind of growth improvement may differ, and should be empirically determined in each particular case.

11.3.2 Pulsed treatment

Pulsed treatment was initially intended to reduce any potential negative effects of clove heating during treatment. However, after comparing the effects of continuous and pulsed pilot-scale E-mode glow plasma, it is clear that the two treatments have an entirely comparable effect on sprout and root growth of unpeeled cloves. While at these particular discharge parameters and treatment times, there is no positive effect on any of the measured parameters, there was also no significant difference between the results of the continuous and pulsed treatments. This confirms the suggestion that it is the O atom dose received by the clove which ultimately determines the biological effect. Further, it suggests that at least at the selected treatment conditions ($P_N = 200$ W, $p = 50$ Pa, $t = 15 - 120$ s, pulsed: 5 s on, 5 s off), the temperature increase of the clove during plasma treatment does not impact sprout and root growth to a degree where pulsed treatment would be of particular benefit.

11.3.3 H-mode afterglow

Differences between the responses to E-mode glow and H-mode afterglow treatments should highlight the importance of certain plasma reactive species for the sprout and root growth improvement of garlic cloves. The growth of unpeeled cloves after treatment in the pilot-scale plasma H-mode afterglow at $P_N = 200$ or 300 W, $p = 50$ s, is considered. Sprout growth is shown in Figure 11.6. While some mean sprout length values considerably exceed those of the untreated control, no clear trend is apparent, and there are no significant differences at any of the treatment times. Root length is shown in Figure 11.7. It does not significantly deviate from the untreated control for either of the treatment conditions. The same is true for root number, shown in Figure 11.8, with the exception that the root number at $P_N = 200$ W, $t = 120$ s is significantly lower than at all of the remaining treatment times at $P_N = 200$ W.

The H-mode afterglow does not significantly affect sprout growth, at least at the conditions used in these experiments. Nonetheless, non-significant sprout length improvements are seen in Figure 11.6. We see a similar effect on root length (Figure 11.7) and number (Figure 11.8), although the non-significant positive effect is primarily an acceleration of root growth, as a comparable plateau is gradually approached by untreated and treated cloves.

The O atom dose received by samples in the H-mode afterglow is generally lower than during exposure to the E-mode glow at the same discharge parameters. For example, as seen in Figure 6.6, the n_O in the afterglow position at $P_N = 300$ W, $p = 30$ Pa, is about $2.5 - 3 \times 10^{21}$ m^{-3}. This is comparable to the n_O at $P_N = 100$ W in the E-mode glow position. It is clear that for the same O atom dose to be received, treatment times in the H-mode afterglow must be longer. We have shown that with suitable treatment times, there are no drastically negative effects on the cloves with E-mode glow treatment. As such, there is no reason to not prefer its use over that of the H-mode afterglow.

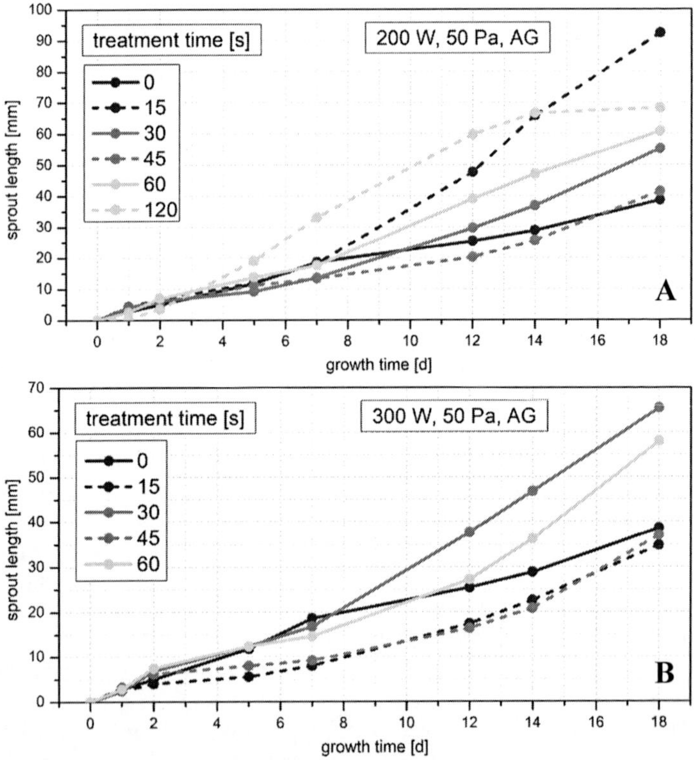

Figure 11.6: Growth of unpeeled clove sprouts, pilot-scale H-mode AG. (A) $P_N = 200$ W; (B) $P_N = 300$ W. $p = 50$ Pa, $t = 0 - 120$ s.

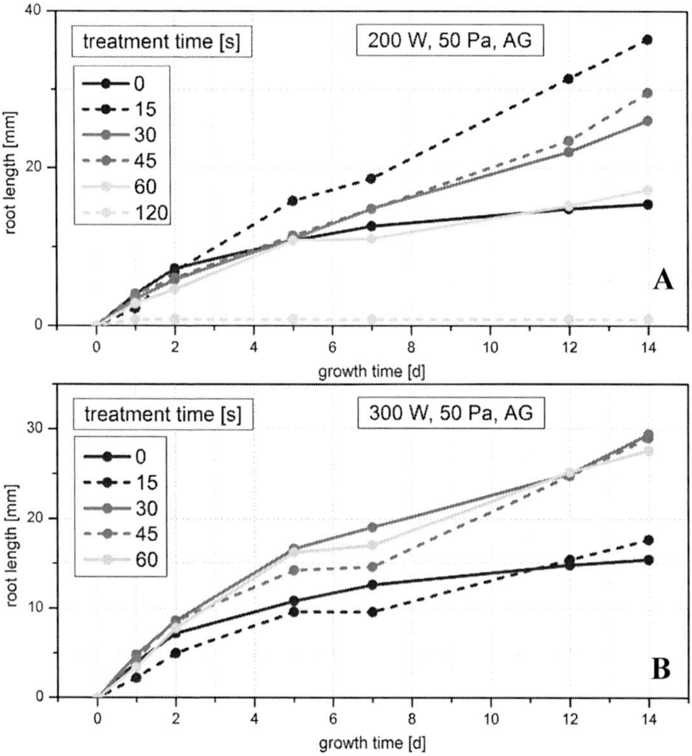

Figure 11.7: Growth of unpeeled clove roots, pilot-scale H-mode AG. (A) $P_N = 200$ W; (B) $P_N = 300$ W. $p = 50$ Pa, $t = 0 – 120$ s.

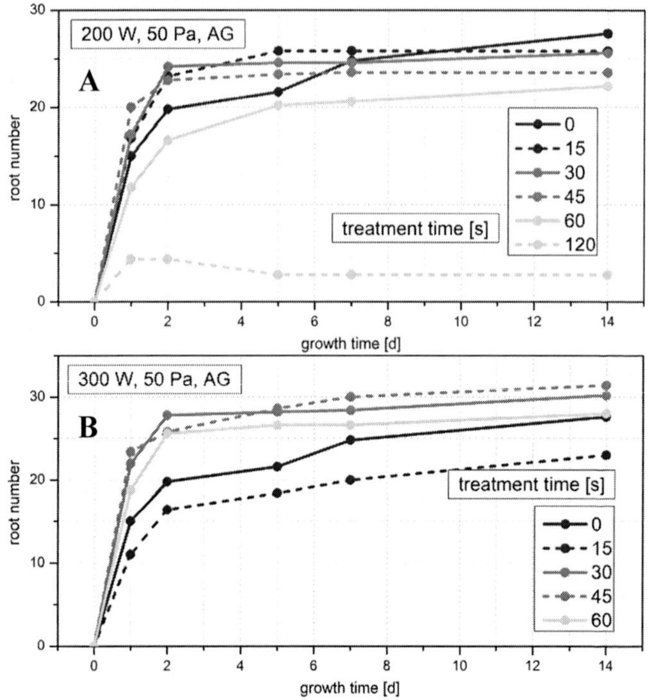

Figure 11.8: Unpeeled clove root number, pilot-scale H-mode AG. (A) $P_N = 200$ W; (B) $P_N = 300$ W. $p = 50$ Pa, $t = 0 – 120$ s.

11.4 Peeled cloves

In parallel to the discussion regarding unpeeled cloves, let us now consider the effects of pilot-scale E-mode glow plasma on the sprout and root growth of peeled garlic cloves. Again, the effect of increasing generator power and treatment time at a constant working pressure is regarded. Sprout growth is seen in Figure 11.9. At a generator power of $P_N = 100$ W, we see no improvement for shorter plasma treatment times, but the sprout length increase is significant for the 90 s and 120 s treatment times from 9 days onward, as well as for the 60 s treatment time from 14 days onwards. Similarly, at $P_N = 200$ W, the sprout length is significantly improved for the 90 s treatment time from 7 days onward, and at $P_N = 300$ W, for the 90 s treatment time from 5 days onward. The treatment times that show no significant improvement nonetheless all exhibit a trend of mean sprout length increase up to $t = 90$ s, followed by a decrease at $t = 120$ s.

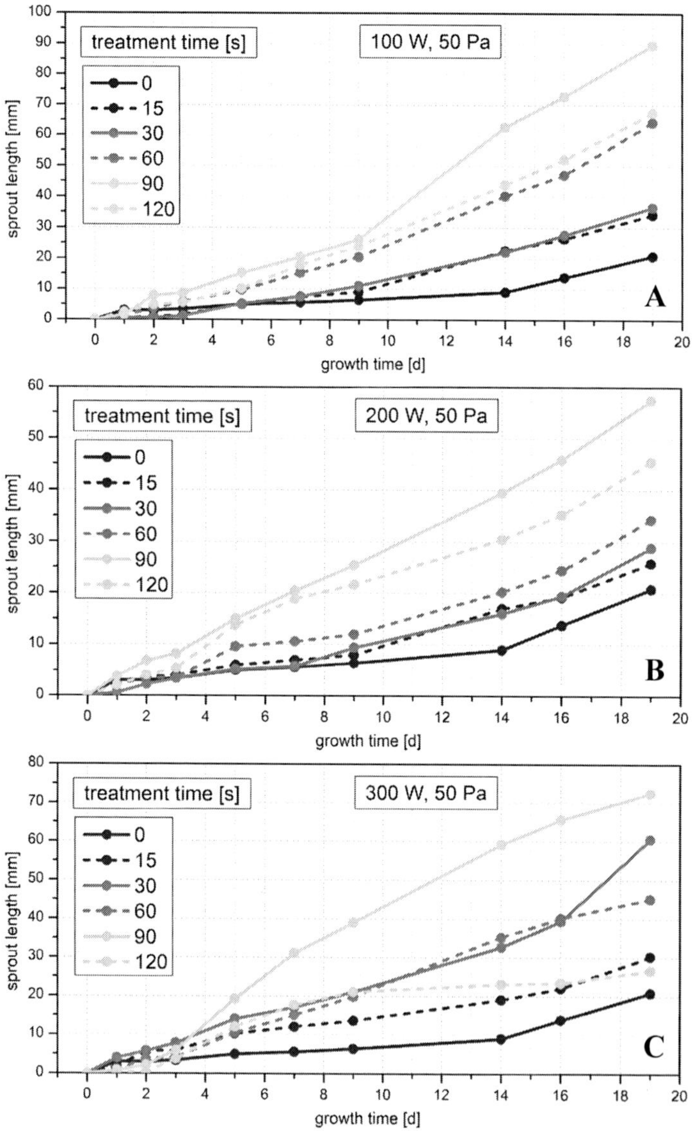

Figure 11.9: Growth of peeled clove sprouts, pilot-scale E-mode glow. (A) P_N = 100 W; (B) P_N = 200 W; (C) P_N = 300 W. p = 50 Pa, t = 0 – 120 s.

It is particularly interesting to note that at all treatment conditions at $p = 50$ Pa, the treatment time to induct significant sprout length improvement is the same: 90 s. In addition, treatment times of 15 – 60 s have a positive effect on sprout growth regardless of P_N, while the longer treatment time of 120 s again diminishes the positive effect, especially at $P_N = 300$ W. This observation is in contrast with the results of unpeeled cloves treatment, where the optimal treatment time varies with P_N, suggesting the importance of the received O atom dose over treatment time.

Here, considering these findings and referring back to Table 11.1, we can again calculate the optimal O atom dose, this time for peeled clove treatment. The results are not as uniform as for unpeeled cloves, as the approximate O atom dose ranges for sprout improvement are $1.4 - 2.8 \times 10^{26}$ (P_N = 100 W), $5 - 6.6 \times 10^{26}$ (P_N = 200 W), and $2.2 - 6.7 \times 10^{26}$ ($P_N = 300$ W). The overlap of these ranges is only partial, suggesting that for sprout growth of peeled cloves, treatment time is a more important parameter than P_N.

In peeled cloves, root growth is affected by plasma in a different pattern than sprout growth. Root length is seen in Figure 11.10. It is not consistently, significantly improved compared to the untreated control at any of the treatment conditions. Nonetheless, the mean root length is affected by treatment time. At each discharge condition, there is a treatment time that has the most pronounced positive effect on mean root length. At a working pressure of 50 Pa, these treatment times are 60 s, 30 s, and 15 s, for P_N = 100 W, 200 W, and 300 W, respectively. We see that this most effective treatment time decreases with increasing P_N (and thus increasing n_O). Below these treatment times, root length is typically comparable to the untreated control, while above, it is increased, but not as much as in the optimal case. The exception is treatment at $P_N = 300$ W, where all treatments longer than 15 s have decreased root length. Here, the O atom dose range for optimal root length improvement, as discerned from Table 11.1, is more uniform across P_N, about $1 - 2 \times 10^{26}$.

Root number, presented in Figure 11.11, is also not consistently significantly improved compared to the untreated control. Nonetheless, the mean root number is affected by treatment time, but there is lack of a clear trend. At P_N = 100 W and 200 W, the optimal treatment time seems to be 60 s, while longer treatment times also positively affect the mean root number. In the case of $P_N = 300$ W, the effect of the treatment times appears to be comparable, with the exception of 120 s, which decreases the root number.

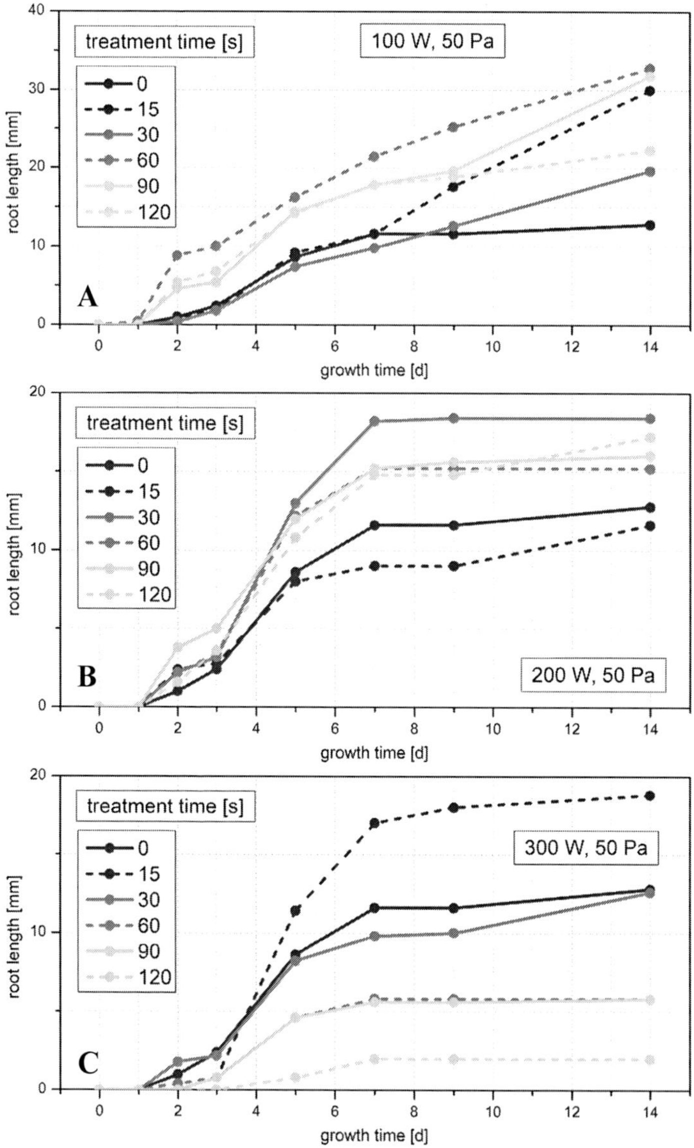

Figure 11.10: Growth of peeled clove roots, pilot-scale E-mode glow. (A) $P_N = 100$ W; (B) $P_N = 200$ W; (C) $P_N = 300$ W. $p = 50$ Pa, $t = 0 - 120$ s.

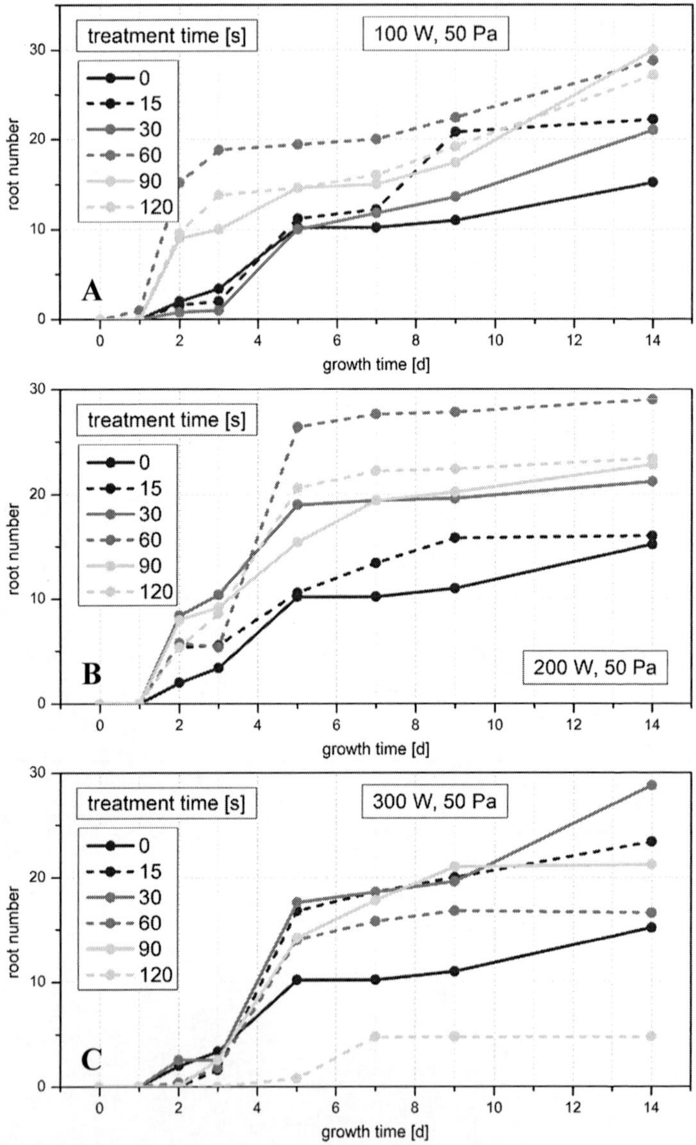

Figure 11.11: Peeled clove root number, pilot-scale E-mode glow. (A) $P_N = 100$ W; (B) $P_N = 200$ W; (C) $P_N = 300$ W. $p = 50$ Pa, $t = 0 - 120$ s.

11.5 Mechanisms of sprout and root growth improvement

11.5.1 Sprout growth

If we compare responses of unpeeled (Figure 11.3) and peeled cloves (Figure 11.9) sprout growth to plasma using the same discharge parameters, we see that their behavior differs. The effect on peeled cloves is more immediate, as we already see a sprout length response at $P_N = 100$ W, which leaves unpeeled cloves unaffected. On the other hand, at higher P_N (200 and 300 W), the optimal response of unpeeled cloves appears at a shorter treatment time, but the improvement is less drastic than for peeled cloves.

Considering all of the above, we propose that with treated unpeeled cloves, the protective leaf acts as a sort of "buffer". On the one hand, it prevents the immediate contact of plasma species with the storage leaf surface, leaving it essentially unaffected. On the other hand, as shown by XPS surface analysis (Table 8.2), water contact angle measurements (Figure 10.1), and water uptake results (Figure 10.5), the treatment causes hydrophilization of the protective leaf surface and accelerates its water uptake. Therefore, it allows for increased, uniform exposure of the clove surface to water taken up by the protective leaf, ensuring a steady moisture supply. Water uptake is the first and essential step required for the start of germination in seeds (Taiz et al. 2014), and if exposure of cloves to plasma facilitates this process, then germination and early growth can be improved by these primarily physico-chemical mechanisms.

On the other hand, when peeled cloves are plasma-treated, the reactive plasma species directly affect the – now unprotected – clove surface. This also causes increased water uptake into the clove, but without the "buffer" function of the protective leaf proposed for unpeeled cloves.

In addition, peeled clove treatment results in potentially direct biological effects: an influence of reactive plasma species on the biochemical pathways of the clove. We know that during plasma treatment, the clove is exposed to a steady flux of ROS. In living beings, ROS are traditionally considered damaging due to their oxidative nature and subsequent damage to the molecular components of the cell. However, it is now widely established that ROS in plants play a dual role, as they are also involved in cellular signaling pathways. Indeed, the concept of an "oxidative window for germination" has been proposed to describe a critical range of ROS concentration in seeds that allows for dormancy release and germination (Bailly et al. 2008, 806-814). During seed storage, ROS levels gradually increase, reducing dormancy (Graeber et al. 2012, 1769-86) by affecting the ratio of plant hormones, particularly abscisic acid and gibberellins, through an influence on the expression of relevant genes (Taiz et al. 2014). Using electron paramagnet resonance, one study suggested a higher radical level in plasma-treated seeds; they noted improved germination, and related this to an acceleration of seed metabolism and stimulation of biochemical and physiological processes (Azharonok et al. 2012). Another group concluded that redox reactions in plasma irradiated radish sprout seeds are influenced by the plasma treatment (Hayashi et al. 2011, 5). We propose that parallel mechanisms positively influence the germination and early growth of plasma-treated garlic cloves.

It is at present unclear how the plasma ROS influence the plant signaling pathways. In seeds, it is suggested that to some degree, plasma species do not only affect the very surface, but can also penetrate the partially permeable seed coat, influencing the seed cells directly (Šerá et al. 2010, 2963-2968). In parallel, we have seen in SEM images that small cracks appear in the cuticle of the treated peeled clove (Figure 9.5), which could allow for ROS to access deeper layers, including epidermal cells, and act as signal transmitters. Indeed, ROS are proposed to act as transmitters of environmental cues during germination. Therefore, it is possible to imagine that ROS produced in oxygen plasma would act on the clove as an environmental signal or stressor, activating its biochemical pathways (Bailly et al. 2008, 806-814).

Viscosity of the cytoplasm, influenced by seed hydration, also plays a role in ROS signaling, enabling mobility of the ROS, allowing them to act as messenger molecules (Bailly et al. 2008, 806-814). In this way, the previously described water uptake improvement also contributes to the activation of biochemical processes in the clove.

11.5.2 Root growth

In a mature garlic clove, 20 – 40 unelongated roots develop in the base of the storage leaf, some of which penetrate through the base even before germination begins (Mann 1952, 195-251). It appears that plasma treatment stimulates, or at least accelerates, the growth of these roots. We suggest that this effect, as with sprout growth above, is due to biochemical activation as a result of plasma treatment, as well as the increased water availability thereafter. The required O atom dose for root growth improvement, as calculated above, is comparable for the treatment of peeled and unpeeled cloves.

As seen from Figure 11.4 and Figure 11.5, root growth is positively affected by plasma earlier than sprout growth, but also the first to be negatively affected if the received O atom dose is too large. From comparing Figure 11.4 and Figure 11.5 with Figure 11.10 and Figure 11.11, we notice that just as the positive effect, the negative effect on the roots of treated unpeeled and peeled cloves is also approximately equal at the same discharge conditions and treatment times. As roots are delicate structures which may already be partially exposed to the environment in a treated clove, an O atom dose that is too large may damage the root tips, including the apical meristem required for root elongation.

In addition, it is apparent from laboratory growth of cloves that the roots are preferential targets for fungal attack. Due to their function of water and nutrient absorption, roots generally lack a cuticle (Taiz et al. 2014), and as such, are more vulnerable to pathogen attack. Any damage to the outer epidermis by the reactive plasma species further increases this susceptibility. As root growth is perhaps the most important factor of plant establishment in the early phases of growth, such damage should necessarily be avoided if improved field growth is to be achieved. Discharge parameters should be adjusted accordingly.

11.5.3 Effect of low pressure

Exposure of cloves to low pressure and oxygen flow alone does not affect their laboratory germination and growth. This was previously shown for agricultural seeds as well, where exposure to vacuum alone was found insufficient for germination improvement (Mildaziene et al. 2018, 1700059).

11.5.4 Effect of EM field

In the case of agricultural seeds, exposure to EM field alone can also stimulate germination (Mildaziene et al. 2016, 536-548; Mildaziene et al. 2018, 1700059), and germination, growth and yield of onion has previously been improved by exposure to magnetic fields (De Souza et al. 2014, 63-69). Contrary to these findings, a control exposure to EM field alone does not significantly affect sprout or root growth of garlic cloves. The EM field may be one of several important components that affect the cloves during plasma treatment, and the results may depend on the characteristics of the applied EM field.

11.5.5 Damage to the cloves

At prolonged treatment times, in particular at the highest discharge power (300 W), the cloves sustain significant damage. This is already evident from the sensory properties of cloves, particularly a glassy appearance and an odor of cooked garlic, both indicating excessive temperature increase. It is further reflected in the sprout and root growth response. With longer treatment times, the sprout length, but especially the root length and number, declines considerably.

This is especially evident when comparing sprout growth of unpeeled and peeled cloves (Figure 11.3 and Figure 11.9). Prolonged treatment at the same discharge conditions affects unpeeled cloves much more severely than peeled cloves; sprout growth is negatively affected at shorter treatment times. This is due to the function of the protective leaf, which confines heat and water to the clove, causing the negative effects of increased temperature to appear earlier due to faster heating (Figure 7.3).

12. FIELD GROWTH AND YIELD

12.1 Field growth of garlic

The ultimate goal of plasma treatment of seed materials is improvement to the plant growth in an agricultural setting. Ideally, this constitutes an improvement to the yield, meaning the agricultural output of the plant. Depending on the nature of the crop, the value that constitutes yield may differ. In case of garlic, yield is primarily the mass of the bulb, which is widely used as a spice or seasoning.

To grow garlic plants from treated and untreated cloves, they need to be planted in the field according to suitable agricultural protocols. Planting time may vary depending on the properties of the specific garlic cultivar. As mentioned, the Slovenian autochthonous cultivar *Ptujski spomladanski* may be planted either in the fall or in the spring for a harvest in the summer.

Any plasma-treated garlic cloves should be planted as soon as possible after plasma treatment. Firstly, the functionalization effect is at least partly reversible due to a process known as aging, wherein the surface oxygen content decreases after treatment (Holc et al. 2016, 325-338), as oxygen is lost from the surface, the newly introduced functional groups re-orientate, and impurities are adsorbed (Modic et al. 2012, 98-104). This effect increases with time after treatment. And secondly, due to the plasma effects on the cuticle surface, the clove is more vulnerable to dehydration, attack by pathogens, and general decay. If the cuticle is partly or fully removed from the surface, the storage and shelf life of the cloves will be reduced (Zhang et al. 2013, 425-432).

To obtain the data presented here, the cloves were grown in a field close to Ptuj, Slovenia, by the staff of the seed company Semenarna Ljubljana, Selection and Testing Center Ptuj. All cloves were planted one to two days after plasma treatment. They were planted either in the fall (October) or in the spring (March) and harvested in July, after approximately nine or four months for fall- or spring-planted cloves, respectively.

12.2 Plant height

Garlic plant height gives information about potential acceleration of plant growth. During plant growth, it may be recorded from the point of plant emergence from the soil to the tip of the longest leaf. At harvest, it may be recorded from the bottom of the bulb to the tip of the longest leaf.

Plant height of garlic plants at harvest is shown in Figure 12.1 and Figure 12.2. The differences that are seen between the mean plant heights are not statistically significant. Here, this is a consequence of recording the plant height at harvest, a time when the foliage leaves are already somewhat dry, and often damaged by the heat and dryness of the summer months. Therefore, such plant height results are not expected to be very informative.

In addition, it should be noted that a garlic clove planted in the field may develop along several paths in response to environmental stimuli (Brewster 2008). An unchanged or even decreased plant height may in fact be desired, as it could indicate that the plant devotes more energy and resources to bulbing than growth of the parts above ground. Still, in several cases of plasma treatment of agricultural seeds of various plants, plant height increase after has later translated to an improvement of yield (Li et al. 2018; Li et al. 2016, 1027-1033; Jiang et al. 2014, 54-58). As such, plant height in the field should not be dismissed, but instead followed throughout the plant growth process.

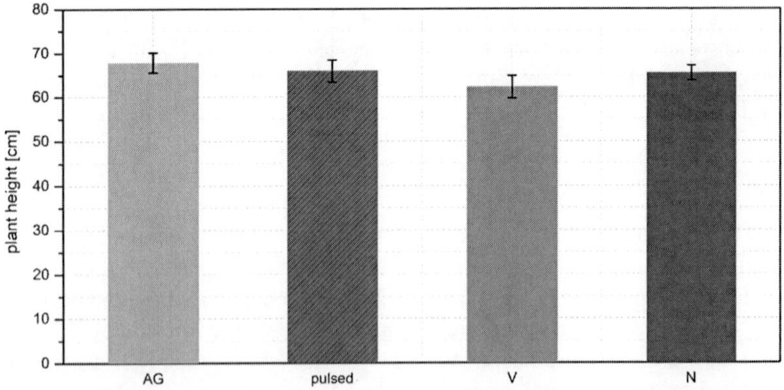

Figure 12.1: Plant height at harvest, pilot-scale plasma. Grown from unpeeled cloves. Planted in October 2017, harvested in July 2018. AG = H-mode afterglow, $P_N = 300$ W, $p = 30$ Pa, $t = 45$ s. Pulsed = E-mode glow, $P_N = 200$ W, $p = 70$ Pa, $t = 30$ s (6 × 5 s on, 5 s off). N = untreated control; V = vacuum control. Error bars represent standard error.

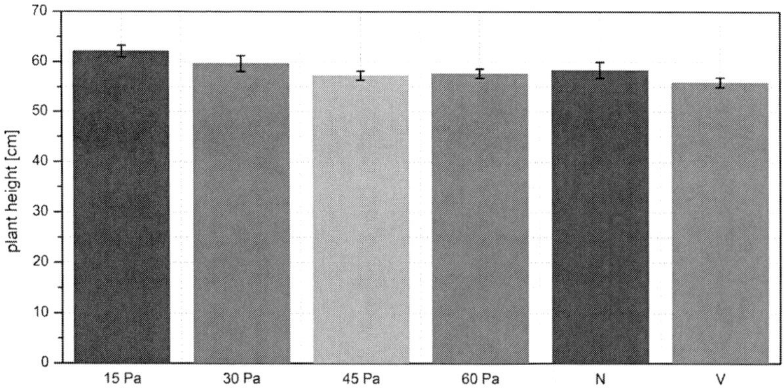

Figure 12.2: Plant height at harvest, industrial-scale plasma. Grown from peeled cloves. Planted in March 2018, harvested in July 2018. P_N = ~ 700 – 900 W, p = 15 – 60 Pa, t = 60 s. N = untreated control; V = vacuum control. Error bars represent standard error.

12.3 Yield

12.3.1 First generation

Yield is the most important agricultural parameter when considering the success of the plasma treatment of seed material. To obtain bulb mass, the harvested garlic should be dried, cleaned, leaves and roots removed, and the mass of each bulb recorded.

First, let us consider yield increase as a consequence of pilot-scale plasma treatment in two different regimens: one H-mode afterglow, one E-mode glow pulsed. Figure 12.3 shows the yield of garlic grown from plasma-treated and untreated unpeeled cloves. Both plasma treatment regimens increase the mean bulb mass compared to the control treatments, although only the increase caused by the afterglow treatment is significant.

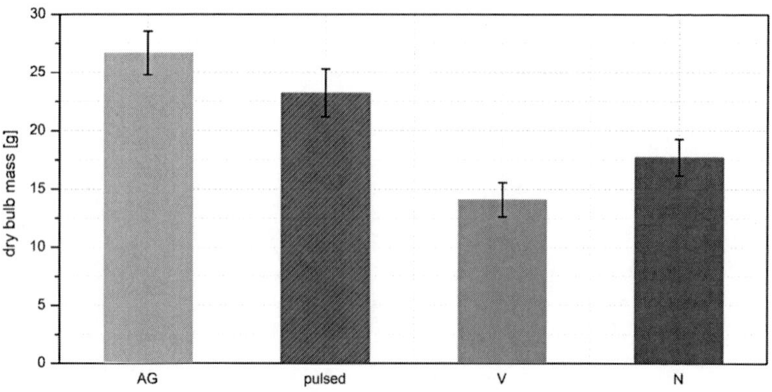

Figure 12.3: Average mass of dried bulb, pilot-scale plasma. Grown from unpeeled cloves. Planted in October 2017, harvested in July 2018. AG = H-mode afterglow, P_N = 300 W, p = 30 Pa, t = 45 s. Pulsed = E-mode glow, P_N = 200 W, p = 70 Pa, t = 30 s (6 × 5 s on, 5 s off). N = untreated control; V = vacuum control. Error bars represent standard error.

Second, in Figure 12.4, let us look at the yield of garlic grown from industrial-scale plasma-treated and untreated peeled cloves. There is an 11 % increase of dried bulb mass for the 15 Pa treatment. Very limited increases are also seen for the 30 Pa and 45 Pa treatments, while the yield after the 60 Pa treatment is decreased. The differences are not statistically significant. As described above, this minimal effect is due to the gentler treatment conditions as a consequence of lower power density. A further factor affecting these results is the time of planting. While the pilot-scale treated cloves from Figure 12.3 were planted in the fall, the industrial-scale planted cloves from Figure 12.4 were planted in the spring. The overwintering of the clove in the soil is important for the garlic plant development, and yields are higher if garlic is planted in the fall (Orlowski et al. 1994, 79-89). The spring-planted garlic has less time for bulb development, and therefore, the differences between the yield of differently treated groups will be less pronounced.

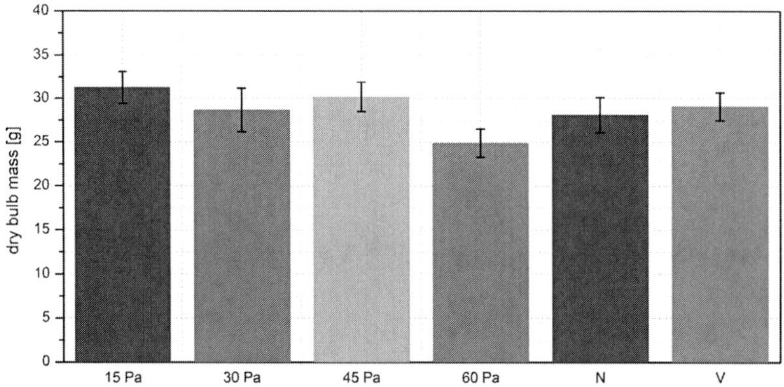

Figure 12.4: Average mass of dried bulb, industrial-scale plasma system. Grown from peeled cloves. Planted in March 2018, harvested in July 2018. P_N = ~ 700 – 900 W, p = 15 – 60 Pa, t = 60 s. N = untreated control; V = vacuum control. Error bars represent standard error.

The mechanism of yield improvement stems from the biological effects of plasma treatment on the clove, as outlined earlier. For the afterglow treatment, the improvements in bulb mass correlate with the improvement in sprout growth. For the pulsed treatment, there is no directly correlated result, but improved sprout growth was recorded in a preliminary experiment. For the industrial-scale treatment, we also see a correlation, as the treatment at 15 Pa caused the highest mean sprout length increase as well as mean bulb mass increase, although none are statistically significant. Parallel findings from plasma treatment of seeds confirm the correlation. Rapeseed seedling height increased after seed plasma treatment, and was accompanied by an increase in yield (Li et al. 2018). In plasma treatment of wheat, medium pressure air/oxygen plasma treatment similarly induced seedling length increase, which translated to a subsequent yield increase (Roy et al. 2017, 13-28). Further, after low-pressure He plasma treatment, wheat yield increase in the field was preceded by increases in plant height and root length at both seedling and booting stage (Jiang et al. 2014, 54-58).

This effect could relate to a general stimulation of plant vigor. One factor is the increased imbibition due to increased hydrophilicity, which, on the one hand, signifies enhanced transmission of water and oxygen (Dobrin et al. 2015, 255-260) as well as nutrients (Zahoranová et al. 2016, 397-414) into the clove. On the other hand, in seeds, imbibition is also essential for the start of physiological activities associated with germination.

Germination is related to the activation of many biochemical pathways within the cells (Ji et al. 2016, 117-28); for example, an increase in chlorophyll content in plants from plasma-treated seeds was recorded (Jiang et al. 2014, 54-58). All these factors can strengthen the plant and improve its growth. Further, root growth improvement is particularly important. As root is the plant organ responsible for nutrient absorption, increased root growth and activity after plasma treatment is linked to increased nutrient uptake (Jiang et al. 2018, 044007), which again improves plant growth. In particular, when garlic is planted in the spring, it is important that the root system is established quickly and efficiently in order for the plant to grow properly.

12.3.2 Second and third generations

To examine potential long-lasting effects of plasma treatment on garlic field growth, the bulbs grown from initially plasma-treated cloves may be stored and planted in the next suitable season without any additional plasma treatment. Here, such cloves were grown as the second or third generation of garlic. Figure 12.5 and Figure 12.6 show dry bulb mass of the 2nd and 3rd generation of garlic grown from cloves treated in plasma before 1st generation planting. No additional treatment took place between the 1st and 2nd generation or between the 2nd and 3rd generation. All garlic was stored, planted and maintained as per the standard procedure.

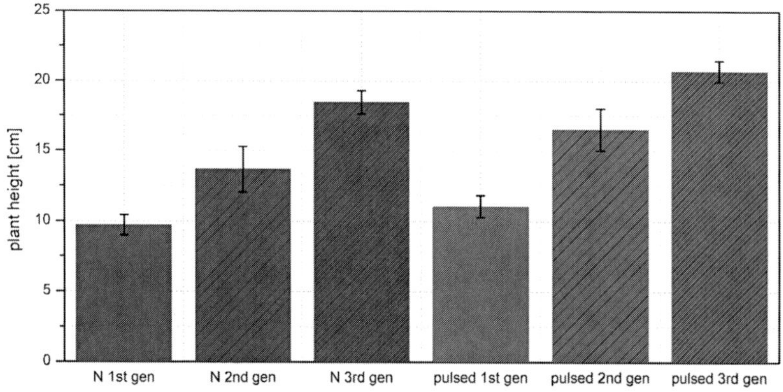

Figure 12.5: Average mass of dried bulb, 2nd and 3rd generation of garlic. Grown from unpeeled cloves, 1st generation planted in May 2016, harvested in July 2016. 2nd generation planted in October 2016, harvested in July 2017. 3rd generation planted in October 2017, harvested in July 2018. Plasma treatment of 1st generation: pilot-scale plasma pulsed E-mode glow, $P_N = 200$ W, $p = 70$ Pa, $t = 30$ s (6×5 s on, 5 s off). Error bars represent standard error.

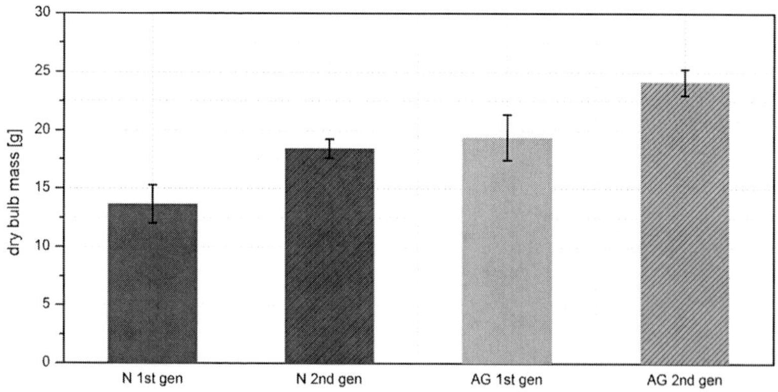

Figure 12.6: Average mass of dried bulb, 2nd generation of garlic. Grown from unpeeled cloves, 1st generation planted in October 2016, harvested in July 2017. 2nd generation planted in October 2017, harvested in July 2018. Plasma treatment of 1st generation: pilot-scale plasma H-mode afterglow, $P_N = 200$ W, $p = 70$ Pa, $t = 45$ s. Error bars represent standard error.

For the garlic treated with the pulsed regimen in the first generation (Figure 12.5), the mean bulb mass increases compared to the bulb mass of untreated garlic. This trend continues through the second and third generation. However, none of the differences between treated and untreated conditions within the same generation are statistically significant. For the garlic treated with the afterglow regimen in the first generation (Figure 12.6), the mean bulb mass also increases compared to the bulb mass of untreated garlic, in both generations. The difference is not statistically significant in the first generation, but is in the second.

These findings indicate that the stimulating effect of plasma treatment on yield is preserved throughout the generations, without additional treatment. No studies so far have considered potential effects of plasma treatment on the yield of the next generation. One study has used plasma on seeds of perennial woody plants to find long-term positive effects on plant morphometric traits, such as stem and root branching or leaf number and surface (Mildaziene et al. 2016, 536-548). This shows that the effects of plasma treatments on plants can be long-lasting. However, our yield results are the first instance indicating that the effects can be carried over into the next generation through non-sexual propagation. We must consider that this effect may simply result from the fact that planting bigger garlic cloves yields bigger bulbs. Even if this is the case, however, it indicates that one-time plasma treatment of cloves can have prolonged stimulating effects on garlic yield.

We see in Figure 12.5 and Figure 12.6 that the mean mass of bulbs from treated and untreated cloves also increases from generation to generation. Potential reasons for this are mainly agricultural and include longer development time (fall planting as opposed to spring planting) or better conditions for garlic growth in that particular year.

CONCLUSIONS

From the findings presented in this book, it is clear that using suitably selected parameters, low pressure inductively coupled RF oxygen plasma can stimulate the growth of garlic clove sprouts and roots, as well as have a positive effect on the yield of garlic grown from such cloves. Further, this yield improvement can persist throughout generations of the garlic plant without the need for additional plasma treatment. This improvement is most likely a consequence of combined physico-chemical and biological action of the reactive plasma species. In cases where the results are not statistically significant, clear trends of improvement are nonetheless often present. The non-significant results are often due to biological variability, which is considerable in garlic cloves, as they may vary in size, shape, water content, or other parameters, even within a single bulb.

There are differences between the physico-chemical and biological responses of peeled and unpeeled cloves to plasma treatment. This is due to a variety of differences between the two. The initial chemical composition of the peeled and unpeeled surface differs, and the peeled clove is more hydrophobic than the unpeeled clove. Further, the protective leaf offers a shielding function to the clove within, and retains at least some of this function during plasma treatment. A larger amount of water is released from the peeled clove during exposure to plasma, which affects the plasma chemistry, making it less oxidizing, and the functionalization and etching processes slower. As a result, the time course of changes to the surface chemistry is different. The surface morphological changes are considerably different as well; whereas the protective leaf is quite easily etched, the reshaping of the peeled clove surface is a more complex process at the same discharge conditions.

All of this also affects sprout and root growth. Sprouts of untreated unpeeled cloves grow more readily due to uniform water exposure. However, the relative stimulating effect on the peeled cloves is larger due to the comparably more direct exposure to reactive plasma species. In addition, during plasma exposure, the temperature of unpeeled cloves increases faster than for peeled cloves, resulting in earlier clove damage. On the other hand, with peeling, the clove loses its protective barrier, making it more vulnerable to negative environmental effects. This includes plasma, which may negatively affect cloves with prolonged exposure. Roots are particularly vulnerable, and the required O atom dose for root growth improvement is lower than for sprout growth improvement. The decision to use peeled or unpeeled cloves should be taken based on processing capabilities, time from treatment to planting, and preliminary tests in the particular plasma system used for treatment.

We propose that mild plasma treatment conditions, akin to the ones used here, only manipulate the garlic clove cuticle and do not negatively affect the epidermal cells, with the exception of the longest treatment times in plasma with high n_O. This is supported by the lack of difference in the fingerprint region of the ATR-FTIR spectra, as well as the lack of nitrogen detection in XPS spectra of etched cloves, which would indicate the presence of proteins from the cells. If etching is extensive enough to damage the cells, the negative effects become apparent due to the clove vulnerability to drying out as well as pathogen attack.

All of the above indicates the presence of an optimal dose of plasma reactive species that a clove should receive to improve sprout growth, root growth and/or yield, as well as avoid damage. This dose can be adjusted by tuning the discharge power, pressure, or treatment time. The selected parameters, in the presence of a sample clove, should also result in an E-mode plasma, where the reactive species density is low enough to not damage the clove. Other plasma configurations may also prove beneficial for growth and yield improvement, as is the case with plasma treatment of agricultural seeds in available literature.

Many of the findings run in parallel with available literature on plasma treatment of agricultural seeds, and can also be applied to that related field of research. Despite biological differences between garlic cloves and other plant seeds, several parallels exist regarding the biological processes of dormancy, germination, and growth.

It is reasonable to propose that the suitable dose of O atoms not only physico-chemically modifies the garlic clove surface, but also affects signaling pathways of plant biochemical processes in a way that stimulates germination and growth, possibly by breaking clove dormancy or affecting its hormonal balance

REFERENCES

Abdel-Fattah, A. F., and Edrees, M. 1972. "A Study on the Composition of Garlic Skins and the Structural Features of the Isolated Pectic Acid." *Journal of the Science of Food and Agriculture* no. 23 (7):871-877. doi: 10.1002/jsfa.2740230708.

Ambrico, P. F., Šimek, M., Morano, M., De Miccolis Angelini, R. M., Minafra, A., Trotti, P., Ambrico, M., Prukner, V., and Faretra, F. 2017. "Reduction of Microbial Contamination and Improvement of Germination of Sweet Basil (Ocimum basilicum L.) Seeds via Surface Dielectric Barrier Discharge." *Journal of Physics D: Applied Physics* no. 50 (30):305401. doi: 10.1088/1361-6463/aa77c8.

Araújo, S. d. S., Paparella, S., Dondi, D., Bentivoglio, A., Carbonera, D., and Balestrazzi, A. 2016. "Physical Methods for Seed Invigoration: Advantages and Challenges in Seed Technology." *Frontiers in Plant Science* no. 7:646. doi: 10.3389/fpls.2016.00646.

Argüello, J. A., Bottini, R., Luna, R., de Bottini, G. A., and Racca, R. W. 1983. "Dormancy in Garlic (Allium sativum L.) cv. Rosado Paraguayo I. Levels of Growth Substances in "Seed Cloves" under Storage." *Plant and Cell Physiology* no. 24 (8):1559-1563. doi: 10.1093/oxfordjournals.pcp.a076679.

Azharonok, V. V., Filatova, I. I., Poklonski, N. A., Lapčuk, N. M., and Gajdarov, A. G. 2012. Electron Paramagnetic Resonance Study of the Efficiency of RF Electromagnetic Field and RF Plasma Seed Pretreatments for Germination Stimulation of Carrot Seeds. Paper read at VII International conference Plasma Physics and Plasma Technology, at Minsk, Belarus.

Bailly, C., El-Maarouf-Bouteau, H., and Corbineau, F. 2008. "From Intracellular Signaling Networks to Cell Death: the Dual Role of Reactive Oxygen Species in Seed Physiology." *Comptes Rendus Biologies* no. 331 (10):806-814. doi: 10.1016/j.crvi.2008.07.022.

Barros, J., Serk, H., Granlund, I., and Pesquet, E. 2015. "The Cell Biology of Lignification in Higher Plants." *Annals of Botany* no. 115 (7):1053-74. doi: 10.1093/aob/mcv046.

Bazelyan, E. M., and Raizer, Y. P. 1997. *Spark Discharge*. 1st ed. Boca Raton: CRC Press.

Bekkara, M. F., Benmimoun, Y., Tilmatine, A., Miloudi, K., and Flazi, S. 2017. "An Effective Approach for Designing a Low Pressure Dc Glow Discharge Plasma Reactor." *Journal of Electrostatics* no. 88:225-231. doi: 10.1016/j.elstat.2017.01.004.

Belmonte, T., Pintassilgo, C. D., Czerwiec, T., Henrion, G., Hody, V., Thiebaut, J. M., and Loureiro, J. 2005. "Oxygen Plasma Surface Interaction in Treatments of Polyolefines." *Surface and Coatings Technology* no. 200 (1–4):26-30. doi: 10.1016/j.surfcoat.2005.02.108.

Bocchini, P., Andalò, C., Pozzi, R., Galletti, G. C., and Antonelli, A. 2001. "Determination of Diallyl Thiosulfinate (Allicin) in Garlic (Allium sativum L.) by High-performance Liquid Chromatography With a Post-Column Photochemical Reactor." *Analytica Chimica Acta* no. 441 (1):37-43. doi: 10.1016/s0003-2670(01)01104-7.

Bol'shakov, A. A., Cruden, B. A., Mogul, R., Rao, M. V. V. S., Sharma, S. P., Khare, B. N., and Meyyappan, M. 2004. "Radio-Frequency Oxygen Plasma as a Sterilization Source." *Aiaa Journal* no. 42 (4):823-832. doi: 10.2514/1.9562.

Bormashenko, E., Grynyov, R., Bormashenko, Y., and Drori, E. 2012. "Cold Radiofrequency Plasma Treatment Modifies Wettability and Germination Speed of Plant Seeds." *Scientific Reports* no. 2:741. doi: 10.1038/srep00741.

Bormashenko, E., Shapira, Y., Grynyov, R., Whyman, G., Bormashenko, Y., and Drori, E. 2015. "Interaction of Cold Radiofrequency Plasma With Seeds of Beans (Phaseolus vulgaris)." *Journal of Experimental Botany* no. 66 (13):4013-4021. doi: 10.1093/jxb/erv206.

Brandenburg, R., Lange, H., von Woedtke, T., Stieber, M., Kindel, E., Ehlbeck, J., and Weltmann, K. D. 2009. "Antimicrobial Effects of UV and VUV Radiation of Nonthermal Plasma Jets." *IEEE Transactions on Plasma Science* no. 37 (6):877-883. doi: 10.1109/tps.2009.2019657.

Brewster, J. L. 2008. *Onions and Other Vegetable Alliums*. 2nd ed. Wallingford: CABI.

Britun, N., Gaillard, M., Ricard, A., Kim, Y. M., Kim, K. S., and Han, J. G. 2007. "Determination of the Vibrational, Rotational and Electron Temperatures in N2 and Ar–n2 RF Discharge." *Journal of Physics D: Applied Physics* no. 40 (4):1022-1029. doi: 10.1088/0022-3727/40/4/016.

Buschhaus, C., and Jetter, R. 2011. "Composition Differences Between Epicuticular and Intracuticular Wax Substructures: How Do Plants Seal Their Epidermal Surfaces?" *Journal of Experimental Botany* no. 62 (3):841-853. doi: 10.1093/jxb/erq366.

Butterbaugh, J. W., Baston, L. D., and Sawin, H. H. 1990. "Measurement and Analysis of Radio Frequency Glow Discharge Electrical Impedance and Network Power Loss." *Journal of Vacuum Science & Technology A: Vacuum, Surfaces, and Films* no. 8 (2):916-923. doi: 10.1116/1.576896.

Caswell, H. L. 1960. "Analysis of the Residual Gases in Several Types of High-Vacuum Evaporators." *IBM Journal of Research and Development* no. 4 (2):130-142. doi: 10.1147/rd.42.0130.

Chen, H. H., Chen, Y. K., and Chang, H. C. 2012. "Evaluation of Physicochemical Properties of Plasma Treated Brown Rice." *Food Chemistry* no. 135 (1):74-79. doi: 10.1016/j.foodchem.2012.04.092.

Clouet, F., and Shi, M. K. 1992. "Interactions of Polymer Model Surfaces With Cold Plasmas: Hexatriacontane as a Model Molecule of High-density Polyethylene and Octadecyl Octadecanoate as a Model of Polyester. I. Degradation Rate Versus Time and Power." *Journal of Applied Polymer Science* no. 46 (11):1955-1966. doi: 10.1002/app.1992.070461108.

Conrads, H., and Schmidt, M. 2000. "Plasma Generation and Plasma Sources." *Plasma Sources Science and Technology* no. 9 (4):441-454. doi: 10.1088/0963-0252/9/4/301.

Cvelbar, U., Krstulović, N., Milošević, S., and Mozetič, M. 2007. "Inductively Coupled RF Oxygen Plasma Characterization by Optical Emission Spectroscopy." *Vacuum* no. 82 (2):224-227. doi: 10.1016/j.vacuum.2007.07.016.

Cvelbar, U., and Mozetič, M. 2007. "Osnove fizike kisikove plazme." *Vakuumist* no. 27 (1):24-33.

da Silva, A. R. M., Farias, M. L., da Silva, D. L. S., Vitoriano, J. O., de Sousa, R. C., and Alves-Junior, C. 2017. "Using Atmospheric Plasma to Increase Wettability, Imbibition and Germination of Physically Dormant Seeds of Mimosa caesalpiniafolia." *Colloids and Surfaces B: Biointerfaces* no. 157:280-285. doi: 10.1016/j.colsurfb.2017.05.063.

De Giorgi, J., Piskurewicz, U., Loubery, S., Utz-Pugin, A., Bailly, C., Mène-Saffrané, L., and Lopez-Molina, L. 2015. "An Endosperm-Associated Cuticle Is Required for Arabidopsis Seed Viability, Dormancy and Early Control of Germination." *PLoS Genetics* no. 11 (12):e1005708. doi: 10.1371/journal.pgen.1005708.

De Souza, A., García, D., Sueiro, L., and Gilart, F. 2014. "Improvement of the Seed Germination, Growth and Yield of Onion Plants by Extremely Low Frequency Non-uniform Magnetic Fields." *Scientia Horticulturae* no. 176:63-69. doi: 10.1016/j.scienta.2014.06.034.

Dhayal, M., Lee, S.-Y., and Park, S.-U. 2006. "Using Low-Pressure Plasma for Carthamus tinctorium L. Seed Surface Modification." *Vacuum* no. 80 (5):499-506. doi: 10.1016/j.vacuum.2005.06.008.

Dobrin, D., Magureanu, M., Mandache, N. B., and Ionita, M.-D. 2015. "The Effect of Non-thermal Plasma Treatment on Wheat Germination and Early Growth." *Innovative Food Science & Emerging Technologies* no. 29:255-260. doi: 10.1016/j.ifset.2015.02.006.

Doliška, A., Vesel, A., Kolar, M., Stana-Kleinschek, K., and Mozetič, M. 2012. "Interaction Between Model Poly(Ethylene Terephthalate) Thin Films and Weakly Ionised Oxygen Plasma." *Surface and Interface Analysis* no. 44 (1):56-61. doi: 10.1002/sia.3769.

Dufour, T., Minnebo, J., Rich, S. A., Neyts, E. C., Bogaerts, A., and Reniers, F. 2014. "Understanding Polyethylene Surface Functionalization by an Atmospheric He/O2 Plasma Through Combined Experiments and Simulations." *Journal of Physics D: Applied Physics* no. 47 (22):224007. doi: 10.1088/0022-3727/47/22/224007.

Egitto, F. D., Vukanovic, V., and Taylor, G. N. 1990. "Plasma Etching of Organic Polymers." In *Plasma Deposition, Treatment, and Etching of Polymers*, edited by d'Agostino, Riccardo, 321-422. San Diego: Academic Press.

El-Fayoumi, I. M., Jones, I. R., and Turner, M. M. 1998. "Hysteresis in the E-to H-mode Transition in a Planar Coil, Inductively Coupled RF Argon Discharge." *Journal of Physics D: Applied Physics* no. 31 (21):3082-3094. doi: 10.1088/0022-3727/31/21/014.

Engeln, R., Mazouffre, S., Vankan, P., Bakker, I., and Schram, D. C. 2002. "Plasma Expansion: Fundamentals and Applications." *Plasma Sources Science and Technology* no. 11 (3A):A100-A104. doi: 10.1088/0963-0252/11/3a/315.

Engineering ToolBox. 2003. "Specific Heat of Food and Foodstuff". Accessed Jun 19th 2019. https://www.engineeringtoolbox.com/specific-heat-capacity-food-d_295.html.

Engineering ToolBox. 2004. "Water - Saturation Pressure ". Accessed May 12th 2019.
https://www.engineeringtoolbox.com/water-vapor-saturation-pressure-d_599.html.

Ensikat, H. J., Boese, M., Mader, W., Barthlott, W., and Koch, K. 2006. "Crystallinity of Plant Epicuticular Waxes: Electron and X-ray Diffraction Studies." *Chemistry and Physics of Lipids* no. 144 (1):45-59. doi: 10.1016/j.chemphyslip.2006.06.016.

Etoh, T. 1986. "Fertility of the Garlic Clones Collected in Soviet Central Asia." *Journal of the Japanese Society for Horticultural Science* no. 55 (3):312-319. doi: 10.2503/jjshs.55.312.

Evert, R. F., and Eichorn, S. E. 2012. *Raven Biology of Plants*. 8th edition ed. New York: W. H. Freeman.

Food and Agriculture Organization of the United Nations. 2015. "FAOSTAT, Browse Data / By Domain". Accessed Dec 15th 2015. http://faostat3.fao.org/browse/Q/QC/E.

Freeman, G. G. 1975. "Distribution of Flavour Components in Onion (Allium cepa L.), Leek (Allium porrum) and Garlic (Allium sativum)." *Journal of the Science of Food and Agriculture* no. 26 (4):471-481. doi: 10.1002/jsfa.2740260413.

Fridman, A. 2008. *Plasma Chemistry*. New York: Cambridge University Press.

Fridman, A., Nester, S., Kennedy, L. A., Saveliev, A., and Mutaf-Yardimci, O. 1999. "Gliding Arc Gas Discharge." *Progress in Energy and Combustion Science* no. 25 (2):211-231. doi: 10.1016/s0360-1285(98)00021-5.

Gerenser, L. J. 1987. "X-Ray Photoemission Study of Plasma Modified Polyethylene Surfaces." *Journal of Adhesion Science and Technology* no. 1 (1):303-318. doi: 10.1163/156856187X00319.

Golda, J., Biskup, B., Layes, V., Winzer, T., and Benedikt, J. 2020. "Vacuum Ultraviolet Spectroscopy of Cold Atmospheric Pressure Plasma Jets." *Plasma Processes and Polymers*:1900216. doi: 10.1002/ppap.201900216.

Gómez-Ramírez, A., López-Santos, C., Cantos, M., García, J. L., Molina, R., Cotrino, J., Espinós, J. P., and González-Elipe, A. R. 2017. "Surface Chemistry and Germination Improvement of Quinoa Seeds Subjected to Plasma Activation." *Scientific Reports* no. 7 (1):5924. doi: 10.1038/s41598-017-06164-5.

Gorjanc, M., Bukošek, V., Gorenšek, M., and Vesel, A. 2009. "The Influence of Water Vapor Plasma Treatment on Specific Properties of Bleached and Mercerized Cotton Fabric." *Textile Research Journal* no. 80 (6):557-567. doi: 10.1177/0040517509348330.

Gorjanc, M., and Mozetič, M. 2014. *Modification of Fibrous Polymers by Gaseous Plasma: Principles, Techniques and Applications*. Saarbrücken: Lambert Academic Publishing.

Graça, J., Schreiber, L., Rodrigues, J., and Pereira, H. 2002. "Glycerol and Glyceryl Esters of ω-Hydroxyacids in Cutins." *Phytochemistry* no. 61 (2):205-215. doi: 10.1016/S0031-9422(02)00212-1.

Graeber, K., Nakabayashi, K., Miatton, E., Leubner-Metzger, G., and Soppe, W. J. J. 2012. "Molecular Mechanisms of Seed Dormancy." *Plant Cell Environ* no. 35 (10):1769-86. doi: 10.1111/j.1365-3040.2012.02542.x.

Grzegorzewski, F., Rohn, S., Kroh, L. W., Geyer, M., and Schlüter, O. 2010. "Surface Morphology and Chemical Composition of Lamb's Lettuce (Valerianella locusta) After Exposure to a Low-pressure Oxygen Plasma." *Food Chemistry* no. 122 (4):1145-1152. doi: 10.1016/j.foodchem.2010.03.104.

Hardin, J. A., Jones, C. L., Weckler, P. R., Maness, N. O., Dillwith, J. W., and Madden, R. D. 2013. "Rapid Quantification of Spinach Leaf Cuticular Wax Using Fourier Transform Infrared Attenuated Total Reflectance Spectroscopy." *Transactions of the Asabe* no. 56 (1):331-339. doi: 10.13031/2013.42579.

Hayashi, N., Nakahigashi, A., Goto, M., Kitazaki, S., Koga, K., and Shiratani, M. 2011. "Redox Characteristics of Thiol Compounds Using Radicals Produced by Water Vapor Radio Frequency Discharge." *Japanese Journal of Applied Physics* no. 50 (8):5. doi: 10.1143/jjap.50.08jf04.

Hayashi, N., Ono, R., Shiratani, M., and Yonesu, A. 2015. "Antioxidative Activity and Growth Regulation of Brassicaceae Induced by Oxygen Radical Irradiation." *Japanese Journal of Applied Physics* no. 54 (6S2):06GD01. doi: 10.7567/jjap.54.06gd01.

Henselová, M., Slováková, Ľ., Martinka, M., and Zahoranová, A. 2012. "Growth, Anatomy and Enzyme Activity Changes in Maize Roots Induced by Treatment of Seeds With Low-temperature Plasma." *Biologia* no. 67 (3):490-497. doi: 10.2478/s11756-012-0046-5.

Hepler, P. K. 2005. "Calcium: a Central Regulator of Plant Growth and Development." *Plant Cell* no. 17 (8):2142-55. doi: 10.1105/tpc.105.032508.

Heredia-Guerrero, J. A., de Lara, R., Domínguez, E., Heredia, A., Benavente, J., and Benítez, J. J. 2012. "Chemical-Physical Characterization of Isolated Plant Cuticles Subjected to Low-Dose Gamma-Irradiation." *Chemistry and Physics of Lipids* no. 165 (8):803-808. doi: 10.1016/j.chemphyslip.2012.10.003.

Hietala, T., Mozes, N., Genet, M. J., Rosenqvist, H., and Laakso, S. 1997. "Surface Lipids and Their Distribution on Willow (Salix) Leaves: a Combined Chemical, Morphological and Physicochemical Study." *Colloids and Surfaces B-Biointerfaces* no. 8 (4-5):205-215. doi: 10.1016/s0927-7765(96)01339-2.

Hoffman, D. M., Singh, B., and Thomas, J. H. 1998. *Handbook of Vacuum Science and Technology*. San Diego: Academic Press.

Holc, M., Junkar, I., Primc, G., Iskra, J., Titan, P., Mlakar, S. G., Kovač, J., and Mozetič, M. 2016. "Improved Sprout Emergence of Garlic Cloves by Plasma Treatment." *Plasma Medicine* no. 6 (3-4):325-338. doi: 10.1615/PlasmaMed.2016019133.

Hutchinson, I. H. 2002. *Principles of Plasma Diagnostics*. 2 ed. Cambridge: Cambridge University Press.

Ji, S. H., Choi, K. H., Pengkit, A., Im, J. S., Kim, J. S., Kim, Y. H., Park, Y., Hong, E. J., Jung, S. K., Choi, E. H., and Park, G. 2016. "Effects of High Voltage Nanosecond Pulsed Plasma and Micro DBD Plasma on Seed Germination, Growth Development and Physiological Activities in Spinach." *Archives of Biochemistry and Biophysics* no. 605:117-28. doi: 10.1016/j.abb.2016.02.028.

Ji, S. H., Ki, S. H., Kang, M. H., Choi, J. S., Park, Y., Oh, J., Kim, S. B., Yoo, S. J., Choi, E. H., and Park, G. 2018. "Characterization of Physical and Biochemical Changes in Plasma Treated Spinach Seed During Germination." *Journal of Physics D: Applied Physics* no. 51 (14):145205. doi: 10.1088/1361-6463/aab2a2.

Ji, S. H., Kim, T., Panngom, K., Hong, Y. J., Pengkit, A., Park, D. H., Kang, M. H., Lee, S. H., Im, J. S., Kim, J. S., Uhm, H. S., Choi, E. H., and Park, G. 2015. "Assessment of the Effects of Nitrogen Plasma and Plasma-Generated Nitric Oxide on Early Development of Coriandum sativum." *Plasma Processes and Polymers* no. 12 (10):1164-1173. doi: 10.1002/ppap.201500021.

Jiang, J., He, X., Li, L., Li, J., Shao, H., Xu, Q., Ye, R., and Dong, Y. 2014. "Effect of Cold Plasma Treatment on Seed Germination and Growth of Wheat." *Plasma Science and Technology* no. 16 (1):54-58. doi: 10.1088/1009-0630/16/1/12.

Jiang, J., Li, J., and Dong, Y. 2018. "Effect of Cold Plasma Treatment on Seedling Growth and Nutrient Absorption of Tomato." *Plasma Science and Technology* no. 20 (4):044007. doi: 10.1088/2058-6272/aaa0bf.

Kamenetsky, R. 2007. "Garlic: Botany and Horticulture." *Horticultural Reviews* no. 33:123-172. doi: 10.1002/9780470168011.ch2.

Kehrer, M., Duchoslav, J., Hinterreiter, A., Mehic, A., Stehrer, T., and Stifter, D. 2020. "Surface Functionalization of Polypropylene Using a Cold Atmospheric Pressure Plasma Jet With Gas Water Mixtures." *Surface and Coatings Technology* no. 384:125170. doi: 10.1016/j.surfcoat.2019.125170.

Kim, J., Katsurai, M., Kim, D., and Ohsaki, H. 2008. "Microwave-Excited Atmospheric-pressure Plasma Jets Using a Microstrip Line." *Applied Physics Letters* no. 93 (19):191505. doi: 10.1063/1.3025841.

Kitazaki, S., Koga, K., Shiratani, M., and Hayashi, N. 2012. "Growth Enhancement of Radish Sprouts Induced by Low Pressure O2 Radio Frequency Discharge Plasma Irradiation." *Japanese Journal of Applied Physics* no. 51 (1S):01AE01. doi: 10.1143/JJAP.51.01AE01.

Koch, K., and Barthlott, W. 2009. "Superhydrophobic and Superhydrophilic Plant Surfaces: an Inspiration for Biomimetic Materials." *Philosophical Transactions of The Royal Society A: Mathematical, Physical and Engineering Sciences* no. 367 (1893):1487-509. doi: 10.1098/rsta.2009.0022.

Koga, K., Thapanut, S., Amano, T., Seo, H., Itagaki, N., Hayashi, N., and Shiratani, M. 2016. "Simple Method of Improving Harvest by Nonthermal Air Plasma Irradiation of Seeds of Arabidopsis thaliana (L.)." *Applied Physics Express* no. 9 (1):3. doi: 10.7567/apex.9.016201.

Kolar, M., and Primc, G. 2016. "Haemostatic Response of Polyethylene Terephthalate Treated by Oxygen and Nitrogen Plasma Afterglows." *International Journal of Polymer Science*. doi: 10.1155/2016/1749285.

Korzec, D., Rapp, J., Theirich, D., and Engemann, J. 1994. "Cleaning of Metal Parts in Oxygen Radio Frequency Plasma: Process study*." *Journal of Vacuum Science & Technology A* no. 12 (2):369-378. doi: 10.1116/1.579249.

Kramida, A., Ralchenko, Y., Reader, J., and NIST ASD Team. 2018. "NIST Atomic Spectra Database (ver. 5.6.1)". Accessed June 10 2019. https://physics.nist.gov/asd.

Kregar, Z., Bišćan, M., Milošević, S., Eleršič, K., Zaplotnik, R., Primc, G., and Cvelbar, U. 2012. "Optical Emission Characterization of Extremely Reactive Oxygen Plasma During Treatment of Graphite Samples." *Materiali in Tehnologije* no. 46 (1):25-30.

Kregar, Z., Milošević, S., and Vesel, A. 2011. "Optical Emission From Oxygen Plasma in E and H Modes." *IEEE Transactions on Plasma Science* no. 39 (11):2502-2503. doi: 10.1109/tps.2011.2160733.

Krongelb, S., and Strandberg, M. W. P. 1959. "Use of Paramagnetic-Resonance Techniques in the Study of Atomic Oxygen Recombinations." *The Journal of Chemical Physics* no. 31 (5):1196-1210. doi: 10.1063/1.1730570.

Kutasi, K., Zaplotnik, R., Primc, G., and Mozetič, M. 2014. "Controlling the Oxygen Species Density Distributions in the Flowing Afterglow of O2/Ar-O2 Surface-Wave Microwave Discharges." *Journal of Physics D: Applied Physics* no. 47 (2). doi: 10.1088/0022-3727/47/2/025203.

Lehocký, M., Drnovská, H., Lapčíková, B., Barros-Timmons, A. M., Trindade, T., Zembala, M., and Lapčík, L. r. 2003. "Plasma Surface Modification of Polyethylene." *Colloids and Surfaces A: Physicochemical and Engineering Aspects* no. 222 (1-3):125-131. doi: 10.1016/s0927-7757(03)00242-5.

Leontiev, R., Hohaus, N., Jacob, C., Gruhlke, M. C. H., and Slusarenko, A. J. 2018. "A Comparison of the Antibacterial and Antifungal Activities of Thiosulfinate Analogues of Allicin." *Scientific Reports* no. 8 (1):6763. doi: 10.1038/s41598-018-25154-9.

Li, L., Jiang, J., Li, J., Shen, M., He, X., Shao, H., and Dong, Y. 2014. "Effects of Cold Plasma Treatment on Seed Germination and Seedling Growth of Soybean." *Scientific Reports* no. 4:5859. doi: 10.1038/srep05859.

Li, L., Li, J., Shao, H., and Dong, Y. 2018. "Effects of Low-vacuum Helium Cold Plasma Treatment on Seed Germination, Plant Growth and Yield of Oilseed Rape." *Plasma Science & Technology* no. 20 (9). doi: 10.1088/2058-6272/aac3d0.

Li, L., Li, J., Shen, M., Hou, J., Shao, H., Dong, Y., and Jiang, J. 2016. "Improving Seed Germination and Peanut Yields by Cold Plasma Treatment." *Plasma Science and Technology* no. 18 (10):1027-1033. doi: 10.1088/1009-0630/18/10/10.

Li, S.-Z., Huang, W.-T., Zhang, J., and Wang, D. 2009. "Optical Diagnosis of an Argon/oxygen Needle Plasma Generated at Atmospheric Pressure." *Applied Physics Letters* no. 94 (11):111501. doi: 10.1063/1.3099339.

LibreTexts. 2019. "Infrared Spectroscopy Absorption Table". Accessed June 30th 2019. https://chem.libretexts.org/Ancillary_Materials/Reference/Reference_Tables/Spectroscopic_Parameters/Infrared_Spectroscopy_Absorption_Table.

Lieberman, M. A., and Lichtenberg, A. J. 2005. *Principles of Plasma Discharges and Materials Processing*. 2nd ed. Hoboken: John Wiley & Sons.

Liu, W., Liu, Y. F., Tao, Y. Q., Yu, Y. J., Jiang, H. M., and Lian, H. Z. 2014. "Comparative Study of Adsorption of Pb(II) on Native Garlic Peel and Mercerized Garlic Peel." *Environmental Science and Pollution Research* no. 21 (3):2054-2063. doi: 10.1007/s11356-013-2112-0.

Luan, P., Kondeti, V. S. S. K., Knoll, A. J., Bruggeman, P. J., and Oehrlein, G. S. 2019. "Effect of Water Vapor on Plasma Processing at Atmospheric Pressure: Polymer Etching and Surface Modification by an

Ar/H2O Plasma Jet." *Journal of Vacuum Science & Technology A* no. 37 (3):031305. doi: 10.1116/1.5092272.

Luo, Y.-R., and Kerr, J. 2012. "Bond Dissociation Energies." In *CRC Handbook of Chemistry and Physics*, edited by Haynes, William M. Boca Raton: CRC Press.

Ma, J. F., Miyake, Y., and Takahashi, E. 2001. "Silicon as a Beneficial Element for Crop Plants." In *Studies in Plant Science*, edited by Datnoff, L. E., Snyder, G. H. and Korndörfer, G. H., 17-39. Elsevier.

Mann, L. K. 1952. "Anatomy of the Garlic Bulb and Factors Affecting Bulb Development." *Hilgardia* no. 21 (8):195-251. doi: 10.3733/hilg.v21n08p195.

Mann, L. K., and Lewis, D. A. 1956. "Rest and Dormancy in Garlic." *Hilgardia* no. 26 (3):161-189. doi: 10.3733/hilg.v26n03p161.

Meng, Y., Qu, G., Wang, T., Sun, Q., Liang, D., and Hu, S. 2017. "Enhancement of Germination and Seedling Growth of Wheat Seed Using Dielectric Barrier Discharge Plasma with Various Gas Sources." *Plasma Chemistry and Plasma Processing* no. 37 (4):1105-1119. doi: 10.1007/s11090-017-9799-5.

Mildaziene, V., Pauzaite, G., Malakauskiene, A., Zukiene, R., Nauciene, Z., Filatova, I., Azharonok, V., and Lyushkevich, V. 2016. "Response of Perennial Woody Plants to Seed Treatment by Electromagnetic Field and Low-temperature Plasma." *Bioelectromagnetics* no. 37:536-548. doi: 10.1002/bem.22003.

Mildaziene, V., Pauzaite, G., Naucienė, Z., Malakauskiene, A., Zukiene, R., Januskaitiene, I., Jakstas, V., Ivanauskas, L., Filatova, I., and Lyushkevich, V. 2018. "Pre-sowing Seed Treatment With Cold Plasma and Electromagnetic Field Increases Secondary Metabolite Content in Purple Coneflower (Echinacea purpurea) Leaves." *Plasma Processes and Polymers* no. 15 (2):1700059. doi: 10.1002/ppap.201700059.

Mirabella, F. M. 1993. *Internal Reflection Spectroscopy: Theory and Applications*. 1st ed. New York: Marcel Dekker.

Misra, N. N., Martynenko, A., Chemat, F., Paniwnyk, L., Barba, F. J., and Jambrak, A. R. 2017. "Thermodynamics, Transport Phenomena, and Electrochemistry of External Field-Assisted Nonthermal Food Technologies." *Critical Reviews in Food Science and Nutrition* no. 58 (11):1832-1863. doi: 10.1080/10408398.2017.1287660.

Modic, M., Junkar, I., Vesel, A., and Mozetič, M. 2012. "Aging of Plasma Treated Surfaces and Their Effects on Platelet Adhesion and Activation." *Surface and Coatings Technology* no. 213:98-104. doi: 10.1016/j.surfcoat.2012.10.026.

Molina, R., López-Santos, C., Gómez-Ramírez, A., Vilchez, A., Espinós, J. P., and González-Elipe, A. R. 2018. "Influence of Irrigation Conditions in the Germination of Plasma Treated Nasturtium Seeds." *Scientific Reports* no. 8. doi: 10.1038/s41598-018-34801-0.

Mondal, S., Rutkoski, J. E., Velu, G., Singh, P. K., Crespo-Herrera, L. A., Guzmán, C., Bhavani, S., Lan, C. X., He, X. Y., and Singh, R. P. 2016. "Harnessing Diversity in Wheat to Enhance Grain Yield, Climate Resilience, Disease and Insect Pest Resistance and Nutrition Through Conventional and Modern Breeding Approaches." *Frontiers in Plant Science* no. 7:991. doi: 10.3389/fpls.2016.00991.

Mozetič, M. 2010. "Surface Modification of Materials Using an Extremely Non-Equilibrium Oxygen Plasma." *Materiali in Tehnologije* no. 44 (4):165-171.

Mozetič, M. 2011. "Extremely Non-equilibrium Oxygen Plasma for Direct Synthesis of Metal Oxide Nanowires on Metallic Substrates." *Journal of Physics D: Applied Physics* no. 44 (17):174028. doi: 10.1088/0022-3727/44/17/174028.

Mozetič, M., Primc, G., Vesel, A., Zaplotnik, R., Modic, M., Junkar, I., Recek, N., Klanjšek-Gunde, M., Guhy, L., Sunkara, M. K., Assensio, M. C., Milošević, S., Lehocky, M., Sedlarik, V., Gorjanc, M., Kutasi, K., and Stana-Kleinschek, K. 2015. "Application of Extremely Non-equilibrium Plasmas in the Processing of Nano and Biomedical Materials." *Plasma Sources Science & Technology* no. 24 (1):015026. doi: 10.1088/0963-0252/24/1/015026.

Mozetič, M., Vesel, A., Stoica, S. D., Vizireanu, S., Dinescu, G., and Zaplotnik, R. 2015. "Oxygen Atom Loss Coefficient of Carbon Nanowalls." *Applied Surface Science* no. 333:207-213. doi: 10.1016/j.apsusc.2015.02.020.

Musil, J., Matouš, J., and Rajský, A. 1993. "Optical Emission Spectra From Microwave Oxygen Plasma Produced by Surfatron Discharge." *Czechoslovak Journal of Physics* no. 43 (5):533-540. doi: 10.1007/bf01589738.

Nafarizal, N., Takada, N., and Sasaki, K. 2008. "Production of Ar Metastable Atoms in the Late Afterglow of Pulse-modulated RF Magnetron Sputtering Plasmas." *Journal of Physics D: Applied Physics* no. 41 (3):035206. doi: 10.1088/0022-3727/41/3/035206.

Nasim, S. A., Dhir, B., Kapoor, R., Fatima, S., Mahmooduzzafar, and Mujib, A. 2010. "Alliin Production in Various Tissues and Organs of Allium Sativum Grown Under Normal and Sulphur-supplemented in Vitro Conditions." *Plant Cell Tissue and Organ Culture* no. 101 (1):59-63. doi: 10.1007/s11240-009-9664-1.

National Institute of Standards and Technology. 2019. "Computational Chemistry Comparison and Benchmark DataBase" (Release 20). Accessed 2020-04-30. https://cccbdb.nist.gov/ie1.asp.

Niemietz, A., Wandelt, K., Barthlott, W., and Koch, K. 2009. "Thermal Evaporation of Multi-component Waxes and Thermally Activated Formation of Nanotubules for Superhydrophobic Surfaces." *Progress in Organic Coatings* no. 66 (3):221-227. doi: 10.1016/j.porgcoat.2009.07.009.

Nishime, T. M. C., Wagner, R., and Kostov, K. G. 2020. "Study of Modified Area of Polymer Samples Exposed to a He Atmospheric Pressure Plasma Jet Using Different Treatment Conditions." *Polymers* no. 12 (5):1028. doi: 10.3390/polym12051028.

Øiseth, S. K., Krozer, A., Kasemo, B., and Lausmaa, J. 2002. "Surface Modification of Spin-coated High-density Polyethylene Films by Argon and Oxygen Glow Discharge Plasma Treatments." *Applied Surface Science* no. 202 (1-2):92-103. doi: 10.1016/s0169-4332(02)00928-5.

Orlowski, M., Rekowska, E., and Dobromilska, R. 1994. "Effect on the Yield of Garlic (Allium sativum L.) of Autumn and Spring Planting Using Different Methods of Seedstalk Trimming." *Folia Horticulturae (Poland)* no. 6 (2):79-89.

Patiño, P., Hernández, F. E., and Rondón, S. 1995. "Reactions of O(3P) with Secondary C-H Bonds of Saturated Hydrocarbons in Nonequilibrium Plasmas." *Plasma Chemistry and Plasma Processing* no. 15 (2):159-171. doi: 10.1007/bf01459694.

Pawłat, J., Starek, A., Sujak, A., Kwiatkowski, M., Terebun, P., and Budzeń, M. 2018. "Effects of Atmospheric Pressure Plasma Generated in GlidArc Reactor on Lavatera thuringiaca L. seeds' Germination." *Plasma Processes and Polymers* no. 15 (2):1700064. doi: 10.1002/ppap.201700064.

Pawłat, J., Starek, A., Sujak, A., Terebun, P., Kwiatkowski, M., Budzen, M., and Andrejko, D. 2018. "Effects of Atmospheric Pressure Plasma Jet Operating with DBD on Lavatera thuringiaca L. seeds' Germination." *PLoS One* no. 13 (4):e0194349. doi: 10.1371/journal.pone.0194349.

Pérez Piza, M. C., Prevosto, L., Zilli, C., Cejas, E., Kelly, H., and Balestrasse, K. 2018. "Effects of Non-thermal Plasmas on Seed-borne Diaporthe/Phomopsis Complex and Germination Parameters of Soybean Seeds." *Innovative Food Science & Emerging Technologies* no. 49:82-91. doi: 10.1016/j.ifset.2018.07.009.

Perkins, M. C., Roberts, C. J., Briggs, D., Davies, M. C., Friedmann, A., Hart, C. A., and Bell, G. A. 2005. "Surface Morphology and Chemistry of Prunus laurocerasus L. Leaves: a Study Using X-Ray Photoelectron

Spectroscopy, Time-of-Flight Secondary-Ion Mass Spectrometry, Atomic-Force Microscopy and Scanning-Electron Microscopy." *Planta* no. 221 (1):123-134. doi: 10.1007/s00425-004-1417-0.

Phan, L., Yoon, S., and Moon, M.-W. 2017. "Plasma-Based Nanostructuring of Polymers: A Review." *Polymers* no. 9 (12):417. doi: 10.3390/polym9090417.

Primc, G., Balat-Pichelin, M., Mozetič, M., and Vesel, A. 2017. "Oxygen Atom Density Within the Interface Between Glowing Oxygen Plasma and Equilibrium Gas." *Vacuum* no. 143:158-164. doi: 10.1016/j.vacuum.2017.06.011.

Primc, G., Zaplotnik, R., Vesel, A., and Mozetic, M. 2011. "Microwave Discharge as a Remote Source of Neutral Oxygen Atoms." *AIP Advances* no. 1 (2):022129. doi: 10.1063/1.3598415.

Puač, N., Gherardi, M., and Shiratani, M. 2018. "Plasma Agriculture: A Rapidly Emerging Field." *Plasma Processes and Polymers* no. 15 (2):1700174. doi: 10.1002/ppap.201700174.

Puliyalil, H., and Cvelbar, U. 2016. "Selective Plasma Etching of Polymeric Substrates for Advanced Applications." *Nanomaterials (Basel)* no. 6 (6):108. doi: 10.3390/nano6060108.

Rabinowitch, H. D., and Brewster, J. L. 1989. *Onions and Allied Crops, Volume III: Biochemistry, Food Science, and Minor Crops*. Boca Raton: CRC Press.

Rahman, M. H., Haque, M. S., Karim, M. A., and Ahmed, M. 2006. "Effects of Gibberellic Acid (GA3) on Breaking Dormancy in Garlic (Allium sativum L.)." *International Journal of Agriculture and Biology* no. 8 (1):63-65.

Randeniya, L. K., and de Groot, G. J. J. B. 2015. "Non-Thermal Plasma Treatment of Agricultural Seeds for Stimulation of Germination, Removal of Surface Contamination and Other Benefits: A Review." *Plasma Processes and Polymers* no. 12 (7):608-623. doi: 10.1002/ppap.201500042.

Rhee, Y., Hlousek-Radojcic, A., Ponsamuel, J., Liu, D. H., and Post-Beittenmiller, D. 1998. "Epicuticular Wax Accumulation and Fatty Acid Elongation Activities Are Induced During Leaf Development of Leeks." *Plant Physiology* no. 116 (3):901-911. doi: 10.1104/pp.116.3.901.

Riederer, M., and Müller, C. 2006. *Biology of the Plant Cuticle*, *Annual Plant Reviews*. Oxford: Blackwell Publishing Ltd.

Roy, N. C., Hasan, M. M., Kabir, A. H., Reza, M. A., Talukder, M. R., and Chowdhury, A. N. 2018. "Atmospheric Pressure Gliding Arc Discharge Plasma Treatments for Improving Germination, Growth and Yield of

Wheat." *Plasma Science & Technology* no. 20 (11). doi: 10.1088/2058-6272/aac647.

Roy, N. C., Hasan, M. M., Talukder, M. R., Hossain, M. D., and Chowdhury, A. N. 2017. "Prospective Applications of Low Frequency Glow Discharge Plasmas on Enhanced Germination, Growth and Yield of Wheat." *Plasma Chemistry and Plasma Processing* no. 38 (1):13-28. doi: 10.1007/s11090-017-9855-1.

Saberi, M., Modarres-Sanavy, S. A. M., Zare, R., and Ghomi, H. 2018. "Amelioration of Photosynthesis and Quality of Wheat under Non-Thermal Radio Frequency Plasma Treatment." *Scientific Reports* no. 8 (1):11655. doi: 10.1038/s41598-018-30200-7.

Schreiber, L. 2005. "Polar Paths of Diffusion Across Plant Cuticles: New Evidence for an Old Hypothesis." *Annals of Botany* no. 95 (7):1069-73. doi: 10.1093/aob/mci122.

Schutze, A., Jeong, J. Y., Babayan, S. E., Jaeyoung, P., Selwyn, G. S., and Hicks, R. F. 1998. "The Atmospheric-pressure Plasma Jet: A Review and Comparison to Other Plasma Sources." *IEEE Transactions on Plasma Science* no. 26 (6):1685-1694. doi: 10.1109/27.747887.

Sprang, N., Theirich, D., and Engemann, J. 1995. "Plasma and Ion Beam Surface Treatment of Polyethylene." *Surface and Coatings Technology* no. 74-75:689-695. doi: 10.1016/0257-8972(95)08340-5.

Srisonphan, S. 2018. "Tuning Surface Wettability through Hot Carrier Initiated Impact Ionization in Cold Plasma." *ACS Appl Mater Interfaces* no. 10 (13):11297-11304. doi: 10.1021/acsami.7b19495.

Stevefelt, J., and Robben, F. 1972. "Spectroscopic Study of the Early Afterglow of a Recombining Helium Plasma." *Physical Review A* no. 5 (3):1502-1515. doi: 10.1103/PhysRevA.5.1502.

Stolárik, T., Henselová, M., Martinka, M., Novák, O., Zahoranová, A., and Černák, M. 2015. "Effect of Low-Temperature Plasma on the Structure of Seeds, Growth and Metabolism of Endogenous Phytohormones in Pea (Pisum sativum L.)." *Plasma Chemistry and Plasma Processing* no. 35 (4):659-676. doi: 10.1007/s11090-015-9627-8.

Suhr, H. 1983. "Application of Nonequilibrium Plasmas in Organic Chemistry." *Plasma Chemistry and Plasma Processing* no. 3 (1):1-61. doi: 10.1007/bf00566027.

Šerá, B., Špatenka, P., Šerý, M., Vrchotová, N., and Hrušková, I. 2010. "Influence of Plasma Treatment on Wheat and Oat Germination and Early Growth." *IEEE Transactions on Plasma Science* no. 38 (10):2963-2968. doi: 10.1109/tps.2010.2060728.

Šourková, H., Primc, G., and Špatenka, P. 2018. "Surface Functionalization of Polyethylene Granules by Treatment with Low-Pressure Air Plasma." *Materials* no. 11 (6). doi: 10.3390/ma11060885.

Štěpánová, V., Slavíček, P., Kelar, J., Prášil, J., Smékal, M., Stupavská, M., Jurmanová, J., and Černák, M. 2018. "Atmospheric Pressure Plasma Treatment of Agricultural Seeds of Cucumber (Cucumis sativus L.) and Pepper (Capsicum annuum L.) with Effect on Reduction of Diseases and Germination Improvement." *Plasma Processes and Polymers* no. 15 (2):1700076. doi: 10.1002/ppap.201700076.

Taiz, L., Zeiger, E., Møller, I. M., and Murphy, A. 2014. *Plant Physiology and Development*. 6th ed. New York: Oxford University Press.

Taylor, G. N., and Wolf, T. M. 1980. "Oxygen Plasma Removal of Thin Polymer Films." *Polymer Engineering and Science* no. 20 (16):1087-1092. doi: 10.1002/pen.760201610.

Thirumdas, R., Kothakota, A., Annapure, U., Siliveru, K., Blundell, R., Gatt, R., and Valdramidis, V. P. 2018. "Plasma Activated Water (PAW): Chemistry, Physico-Chemical Pproperties, Applications in Food and Agriculture." *Trends in Food Science & Technology* no. 77:21-31. doi: 10.1016/j.tifs.2018.05.007.

Thirumdas, R., Sarangapani, C., and Annapure, U. S. 2015. "Cold Plasma: A Novel Non-Thermal Technology for Food Processing." *Food Biophysics* no. 10 (1):1-11. doi: 10.1007/s11483-014-9382-z.

Toneli, D. A., Pessoa, R. S., Roberto, M., and Gudmundsson, J. T. 2015. "On the Formation and Annihilation of the Singlet Molecular Metastables in an Oxygen Discharge." *Journal of Physics D: Applied Physics* no. 48 (32):325202. doi: 10.1088/0022-3727/48/32/325202.

Tong, J., He, R., Zhang, X., Zhan, R., Chen, W., and Yang, S. 2014. "Effects of Atmospheric Pressure Air Plasma Pretreatment on the Seed Germination and Early Growth of Andrographis paniculata." *Plasma Science and Technology* no. 16 (3):260. doi: 10.1088/1009-0630/16/3/16.

Uchida, H., Tanabe, K., Nojiri, Y., Haraguchi, H., and Fuwa, K. 1980. "Measurement of Metastable Argon in an Inductively Coupled Argon Plasma by Atomic Absorption Spectroscopy." *Spectrochimica Acta Part B: Atomic Spectroscopy* no. 35 (11-12):881-883. doi: 10.1016/0584-8547(80)80026-7.

Vesel, A., and Mozetič, M. 2001. "Behaviour of Catalytic Probe During Surface Activation of Polyether Sulphone." *Vacuum* no. 61 (2-4):373-377. doi: 10.1016/s0042-207x(01)00147-6.

Vesel, A., and Mozetič, M. 2017. "New Developments in Surface Functionalization of Polymers Using Controlled Plasma Treatments." *Journal of Physics D: Applied Physics*. doi: 10.1088/1361-6463/aa748a.

Vesel, A., Mozetič, M., Hladnik, A., Dolenc, J., Zule, J., Milošević, S., Krstulović, N., Klanjšek-Gunde, M., and Hauptmann, N. 2007. "Modification of Ink-Jet Paper by Oxygen-Plasma Treatment." *Journal of Physics D: Applied Physics* no. 40 (12):3689. doi: 10.1088/0022-3727/40/12/022.

Vesel, A., Zaplotnik, R., Kovač, J., and Mozetič, M. 2018. "Initial Stages in Functionalization of Polystyrene Upon Treatment With Oxygen-plasma Late Flowing Afterglow." *Plasma Sources Science & Technology* no. 27 (9):094005. doi: 10.1088/1361-6595/aad486.

Vialle, M., Touzeau, M., Gousset, G., and Ferreira, C. M. 1991. "Kinetics of O(1S) and O(1D) Metastable Atoms in a DC Oxygen Glow Discharge." *Journal of Physics D: Applied Physics* no. 24 (3):301-308. doi: 10.1088/0022-3727/24/3/011.

Vukušić, T., Vesel, A., Holc, M., Ščetar, M., Režek Jambrak, A., and Mozetič, M. 2018. "Modification of Physico-Chemical Properties of Acryl-Coated Polypropylene Foils for Food Packaging by Reactive Particles from Oxygen Plasma." *Materials* no. 11 (3). doi: 10.3390/ma11030372.

Wang, X. Q., Zhou, R. W., Groot, G., Bazaka, K., Murphy, A. B., and Ostrikov, K. K. 2017. "Spectral Characteristics of Cotton Seeds Treated by a Dielectric Barrier Discharge Plasma." *Scientific Reports* no. 7 (1):5601. doi: 10.1038/s41598-017-04963-4.

Wheale, S. H., Barker, C. P., and Badyal, J. P. S. 1998. "Chemical Reaction Pathways at the Plasma–Polymer Interface." *Langmuir* no. 14 (23):6699-6704. doi: 10.1021/la980759c.

Wickramanayaka, S., Hosokawa, N., and Hatanaka, Y. 1991. "Variation of the Recombination Coefficient of Atomic Oxygen on Pyrex Glass with Applied RF Power." *Japanese Journal of Applied Physics* no. 30 (11A):2897-2900. doi: 10.1143/jjap.30.2897.

Wood, B. J., and Wise, H. 1962. "The Kinetics of Hydrogen Atom Recombination on Pyrex Glass and Fused Quartz." *The Journal of Physical Chemistry* no. 66 (6):1049-1053. doi: 10.1021/j100812a019.

Wu, C., Wang, M., Dong, Y., Cheng, Z., and Meng, H. 2015. "Growth, Bolting and Yield of Garlic (Allium sativum L.) in Response to Clove Chilling Treatment." *Scientia Horticulturae* no. 194:43-52. doi: 10.1016/j.scienta.2015.07.018.

Yin, M., Huang, M., Ma, B., and Ma, T. 2005. "Stimulating Effects of Seed Treatment by Magnetized Plasma on Tomato Growth and Yield." *Plasma Science and Technology* no. 7 (6):3143-3147. doi: 10.1088/1009-0630/7/6/017.

Yoon, S., and Lee, T. G. 2018. "Biological Tissue Sample Preparation for Time-of-Flight Secondary Ion Mass Spectrometry (ToF-SIMS) Imaging." *Nano Convergence* no. 5 (1):24. doi: 10.1186/s40580-018-0157-y.

Young, R. A., Black, G., and Slanger, T. G. 1968. "Reaction and Deactivation of O(1D)." *The Journal of Chemical Physics* no. 49 (11):4758-4768. doi: 10.1063/1.1669957.

Zahoranová, A., Henselová, M., Hudecová, D., Kaliňaková, B., Kováčik, D., Medvecká, V., and Černák, M. 2016. "Effect of Cold Atmospheric Pressure Plasma on the Wheat Seedlings Vigor and on the Inactivation of Microorganisms on the Seeds Surface." *Plasma Chemistry and Plasma Processing* no. 36 (2):397-414. doi: 10.1007/s11090-015-9684-z.

Zahoranová, A., Hoppanová, L., Šimončicová, J., Tučeková, Z., Medvecká, V., Hudecová, D., Kaliňáková, B., Kováčik, D., and Černák, M. 2018. "Effect of Cold Atmospheric Pressure Plasma on Maize Seeds: Enhancement of Seedlings Growth and Surface Microorganisms Inactivation." *Plasma Chemistry and Plasma Processing* no. 38 (5):969-988. doi: 10.1007/s11090-018-9913-3.

Zaplotnik, R., Vesel, A., and Mozetič, M. 2011. "Transition from E to H Mode in Inductively Coupled Oxygen Plasma: Hysteresis and the Behaviour of Oxygen Atom Density." *EPL (Europhysics Letters)* no. 95 (5):55001. doi: 10.1209/0295-5075/95/55001.

Zaplotnik, R., Vesel, A., and Mozetič, M. 2012. "A Fiber Optic Catalytic Sensor for Neutral Atom Measurements in Oxygen Plasma." *Sensors* no. 12 (4):3857-67. doi: 10.3390/s120403857.

Zaplotnik, R., Vesel, A., and Mozetič, M. 2018. "Atomic Oxygen and Hydrogen Loss Coefficient on Functionalized Polyethylene Terephthalate, Polystyrene, and Polytetrafluoroethylene Polymers." *Plasma Processes and Polymers* no. 15 (9):1800021. doi: 10.1002/ppap.201800021.

Zaplotnik, R., Vesel, A., Primc, G., Liu, X., Chen, K., Wei, C., Xu, K., and Mozetič, M. 2016. "Rapid Hydrophilization of Model Polyurethane/Urea (PURPEG) Polymer Scaffolds Using Oxygen Plasma Treatment." *Polymers* no. 8 (4):144. doi: 10.3390/polym8040144.

Zhang, B., Li, R. H., and Yan, J. C. 2018. "Study on Activation and Improvement of Crop Seeds by the Application of Plasma Treating Seeds Equipment." *Archives of Biochemistry and Biophysics* no. 655:37-42. doi: 10.1016/j.abb.2018.08.004.

Zhang, M., Oh, J. K., Cisneros-Zevallos, L., and Akbulut, M. 2013. "Bactericidal Effects of Nonthermal Low-pressure Oxygen Plasma on S. typhimurium LT2 Attached to Fresh Produce Surfaces." *Journal of Food Engineering* no. 119 (3):425-432. doi: 10.1016/j.jfoodeng.2013.05.045.

Zhang, S., van Gaens, W., van Gessel, B., Hofmann, S., van Veldhuizen, E., Bogaerts, A., and Bruggeman, P. 2013. "Spatially Resolved Ozone Densities and Gas Temperatures in a Time Modulated RF Driven Atmospheric Pressure Plasma Jet: An Analysis of the Production and Destruction Mechanisms." *Journal of Physics D: Applied Physics* no. 46 (20):205202. doi: 10.1088/0022-3727/46/20/205202.

Zhao, Y., Li, W., Liu, Z., Liu, J., Zhu, L., Liu, X., and Huang, K. 2018. "Renewable Tb/Eu-Loaded Garlic Peels for Enhanced Adsorption of Enrofloxacin: Kinetics, Isotherms, Thermodynamics, and Mechanism." *ACS Sustainable Chemistry & Engineering* no. 6 (11):15264-15272. doi: 10.1021/acssuschemeng.8b03739.

Zhou, Z., Huang, Y., Yang, S., and Chen, W. 2011. "Introduction of a New Atmospheric Pressure Plasma Device and Application on Tomato Seeds." *Agricultural Sciences* no. 02 (01):23-27. doi: 10.4236/as.2011.21004.